ASPECTS *of*
SPONGE BIOLOGY

Academic Press Rapid Manuscript Reproduction

ASPECTS *of* SPONGE BIOLOGY

Edited by
FREDERICK W. HARRISON

Department of Anatomy
Albany Medical College
Albany, New York

and
RONALD R. COWDEN

School of Medicine
East Tennessee State University
Johnson City, Tennessee

ACADEMIC PRESS   **New York**   **San Francisco**   **London**   **1976**
A Subsidiary of Harcourt Brace Jovanovich, Publishers

ACADEMIC PRESS, INC.
111 Fifth Avenue, New York, New York 10003

United Kingdom Edition published by
ACADEMIC PRESS, INC. (LONDON) LTD.
24/28 Oval Road, London NW1

Library of Congress Cataloging in Publication Data
Main entry under title:

Aspects of sponge biology.

"Developed from a symposium held in Albany,
New York, in May 1975 . . . sponsored by the Society
for Developmental Biology and the Department of
Anatomy, Albany Medical College."
    Includes bibliographical references and index.
    1.   Sponges–Congresses.   I.   Harrison, Frederick
Williams, (date)   II.   Cowden, Ronald R.
III.   Society for Developmental Biology.   IV.   Al-
bany Medical College, Albany, N.Y. Dept. of
Anatomy.
QL371.H86   593'.4   76-3445
ISBN 0–12–327950–X

To Jeffrey D. Rude
who died October 12, 1975 at McMurdo
Sound, Antartica while conducting
research on the biology of sponges.

Contents

List of Contributors and Participants in the General Discussion

Leo W. Buss, Department of Earth and Planetary Sciences, The Johns Hopkins University, Baltimore, Maryland 21218 and Discovery Bay Marine Laboratory, Discovery Bay, St. Anns, Jamaica, W.I.

Jack T. Cecil, Osborn Laboratories of Marine Sciences, New York Aquarium, New York Zoological Society, Brooklyn, New York 11224

Wen-Tien Chen, Department of Biology, Yale University, New Haven, Connecticut 06520

Ronald R. Cowden, School of Medicine, East Tennessee State University, Johnson City, Tennessee 37601

Paul K. Dayton, Scripps Institution of Oceanography, La Jolla, California 92307

George DeNagy, Department of Biological Sciences, Dartmouth College, Hanover, New Hampshire 03755

Paul E. Fell, Department of Zoology, Connecticut College, New London, Connecticut 06320

Thomas M. Frost, Department of Biological Sciences, Dartmouth College, Hanover, New Hampshire 03755

Samuel L.H. Fuller, Department of Limnology, The Academy of Natural Sciences of Philadelphia, Philadelphia, Pennsylvania 19103

Thomas F. Goreau, Marine Sciences Research Center, State University of New York at Stony Brook, Stony Brook, New York 11790 and Discovery Bay Marine Laboratory, Discovery Bay, St. Anns, Jamaica, W.I.

Michael J. Greenberg, Department of Biological Sciences, Florida State University, Tallahassee, Florida and the Marine Biological Laboratory, Woods Hole, Massachusetts

Frederick W. Harrison, Department of Anatomy, Albany Medical College, Albany, New York 12208

Willard D. Hartman, Peabody Museum of Natural History and Department of Biology, Yale University, New Haven, Connecticut 06520

George J. Hechtel, Department of Ecology and Evolution, State University of New York at Stony Brook, Stony Brook, New York 11794

Carter Litchfield, Department of Biochemistry, Rutgers University, New Brunswick, New Jersey 08903

Frank J. Little, Box 692, Clarkson, New York 14430

Reginald W. Morales, Department of Biochemistry, Rutgers University, New Brunswick, New Jersey 08903

Ross F. Nigrelli, Osborn Laboratories of Marine Sciences, New York Aquarium, New York Zoological Society, Brooklyn, New York 11224

Sidney K. Pierce, Jr.,[1] Department of Biological Sciences, Florida State University, Tallahassee, Florida 32306 and the Marine Biological Laboratory, Woods Hole, Massachusetts

Michael A. Poirrier, Department of Biological Sciences, University of New Orleans, New Orleans, Louisiana 70122

Shirley A. Pomponi, University of Miami, Rosensteil School of Marine and Atmospheric Science, 4600 Rickenbacker Causeway, Miami, Florida 33149

Henry M. Reiswig, Redpath Museum and Department of Biology, McGill University, P.O. Box 6070, Montreal, Quebec, Canada

Charlene Reed, Department of Biological Sciences, Florida State University, Tallahassee, Florida 32306 and the Marine Biological Laboratory, Woods Hole, Massachusetts

Vincent H. Resh,[2] Department of Biology, Ball State University, Muncie, Indiana, 47306

Gideon A. Rodan, School of Medicine and Dental Medicine, University of Connecticut, Farmington, Connecticut 06032

Jeffrey D. Rude, Scripps Institution of Oceanography, La Jolla, California 92037

Klaus Ruetzler, Department of Invertebrate Zoology (Lower Invertebrates), National Museum of Natural History, Smithsonian Institution, Washington, D.C. 20560

George D. Ruggieri, Osborn Laboratories of Marine Sciences, New York Aquarium, New York Zoological Society, Brooklyn, New York 11224

Tracy L. Simpson, Department of Biology, University of Hartford, West Hartford, Connecticut 06117

Martin F. Stempien, Jr., Osborn Laboratories of Marine Sciences, New York Aquarium, New York Zoological Society, Brooklyn, New York 11224

John F. Storr, Department of Biology, State University of New York at Buffalo, Buffalo, New York 14214

Felix Wiedenmayer, Naturhistorisches Museum Basel, Basel, Switzerland.

[1] Present Address: Department of Zoology, University of Maryland, College Park, Maryland 20742

[2] Present Address: Division of Entomology and Parasitology, University of California, Berkeley, California 94720

Preface

Aspects of Sponge Biology developed from a symposium held in Albany, New York, in May 1975, sponsored by the Society for Developmental Biology and the Department of Anatomy, Albany Medical College. This symposium brought together the majority of North American investigators of sponge biology. The book is unusual in that, in addition to presentations in current investigations, it contains the symposium participants' discussion of several of the problem areas of sponge biology. The introductory chapter is intended for established investigators in other fields who either wish to study the sponges *per se* or to utilize these animals as model systems to clarify basic biological problems. This book, then, attempts to present the sponges as a challenging, virtually untapped resource for future studies. It includes the most current research in the field yet, simultaneously, leads investigators into research opportunities seen for the near future. *Aspects of Sponge Biology* should prove valuable to invertebrate zoologist, cell and developmental biologists, aquatic biologists, ecologists, investigators of cell surface phenomena, comparative physiologists, and to anyone involved in problems of water quality. We feel that the study of sponge biology is entering into an extremely exciting and rapidly evolving period in which the utilization of techniques unavailable in the recent past will not only provide answers to many of the problems now existing in sponge biology but will also raise challenging new questions.

Fredrick W. Harrison
Ronald R. Cowden

Acknowledgments

We wish to express our appreciation to the Society for Developmental Biology and to the Department of Anatomy, Albany Medical College for their generous support of the "Symposium on Sponge Biology" held at Albany, New York. We are grateful to Dr. Richard G. Skalko, Department of Anatomy, Albany Medical College and Birth Defects Institute, New York State Department of Health, who as Acting Chairman, Department of Anatomy, provided secretarial support and considerable encouragement. Without the conscientious efforts of our typists and technicians, Mrs. Robyn Hymen, Mrs. Rita Brooks, Mrs. Marie Baker, Mrs. Patricia Hicks, and Miss Lori Johnston, meeting publication deadlines would have been impossible.

Introduction and General Discussion

INTRODUCTION: PRINCIPLES AND PERSPECTIVES
IN SPONGE BIOLOGY

by

Frederick W. Harrison
Department of Anatomy
Albany Medical College
Albany, New York   12208

and

Tracy L. Simpson
Department of Biology
University of Hartford
West Hartford, Connecticut   06117

Within recent years, the utilization of sponges in research has increased to a tremendous extent. The application of newer technologies to problems of sponge biology has
greatly clarified many of the problems that plagued earlier
investigators. We feel, however, that the sponges present a
virtually untapped tool for use in basic research with many
areas of utilization and investigation unrecognized. This
chapter is intended, then, to serve as a guide to either the
young investigator beginning a career or the established
scientist who wishes to utilize, for the first time, sponges
as a research vehicle. We realize that it is not possible
or reasonable to review all the different facets of sponge
biology. However, we wish to introduce the novice to: methods for collecting, laboratory maintenance and examination;
key references, including monographs and review papers; and
selected areas of research on sponges which need to be undertaken or need to be reevaluated in light of new techniques
and/or ideas.

Collecting Techniques

In collecting sponges, the methods employed will often be
determined by the collectors' proposed use of the material.
If one is collecting for taxonomic studies of marine sponges,
for instance, it is best, if feasible, to collect entire
specimens. In this case, color photographs, underwater if
possible, of the specimens are quite helpful in species determinations. Conversely, with freshwater sponges, the collection of entire specimens is not essential but one should
collect gemmule-bearing specimens if possible. This is because most taxonomic schemes in use today employ gemmule and

3

gemmosclere (gemmule spicule) morphology as diagnostic criteria. In either case, it is important to separate small,
five cubic millimeter, pieces of sponge into a histological
fixative. Fixatives of preference are Bouins or ethanol-
acetic acid (3:1). Following fixation, specimens should be
processed routinely through washes, etc., with storage in
70% ethanol. The remainder of the specimen may be placed
directly into 70% ethanol or retained as a dried specimen.

Maintenance of Sponges in the Laboratory

The problems of maintenance of marine and freshwater
sponges in the laboratory have been reviewed by Fell ('67).
Although freshwater sponges are notoriously difficult to
maintain in the laboratory, Imlay and Paige ('72) described
a simple laboratory system with continuous water flow in
which freshwater sponges not only survived for three months
but, in most cases, exhibited considerable growth. The continuous flow multichamber system described by these authors
used trout fry food (Glencoe starter granules) fed into the
first prechamber at the rate of ½ gram of feed per day. Although the exact system described, i.e., direct introduction
of raw habitat water, would be impractical for most laboratories, a recirculating system could be easily devised. The
design for an inexpensive recirculating system, in this case
a refrigerated seawater system for marine organisms, was described by Bakus ('65). This system could be easily adapted
for maintenance of freshwater sponges according to the technique of Imlay and Paige ('72).

The various techniques used in laboratory examination of
sponges, i.e. explants, dissociation and reaggregation,
growth of sponges from larvae, production of sponges from
gemmules, and cell culture, have been thoroughly reviewed by
Fell ('67) in a particularly informative article.

Current Problems in Systematics

Until recently, the systematics of freshwater sponges was
hopelessly confused. The revisionary work of Penney and
Racek ('68) brought some degree of order into this chaotic
area and, in particular, demonstrated global evolutionary
patterns within the gemmule-forming spongillids. However,
there are still major areas requiring clarification in
freshwater sponge systematics. Traditionally, skeletal and/
or gemmule morphology have been the basic criteria utilized
in systematic analyses of both freshwater and marine sponges.
Increased recognition of the problems caused by ecomorphic
variation in skeletal and gemmule structure of spongillids
(see Poirrier'69, '74, and this volume) necessitate a more

4

comprehensive approach. Although non-skeletal characteristics have been utilized in taxonomic studies of marine
sponges (see Pomponi, in this volume for a review), their
application as diagnostic criteria in freshwater sponge systematics has been limited. Harrison ('71) used cytochemical
characteristics in defining one species of spongillid and
Arceneaux ('73) applied electrophoretic techniques to
problems in freshwater sponge taxonomy, but there has been
no significant advance in this area.

Especially in widely distributed freshwater species such
as Spongilla lacustris and Eunapius fragilis, possible speciation trends in distant populations should be considered.
For example, as Penney and Racek ('68) noted, the majority of
sub-artic or cold-temperate forms of S. lacustris show morphological characteristics different from those of more
southern forms. While such criteria as the presence or absence of gemmule pneumatic coats (see Poirrier, '69) may reflect ecomorphic variation, there appear to be significant
differences in life history, growth forms, etc., in distant
populations. As discussions in this volume indicate, speciation trends and the question of subspecific status provide
an intriguing area for future research in sponge systematics.

A number of freshwater sponges do not form gemmules and,
thus, present special taxonomic problems. Evaluation of the
evolutionary relationships of these non-spongillid freshwater
sponges has involved the utilization of a number of non-
skeletal characters, particularly developmental characteristics (see Brien, '67a, '67b, '70).

It has now become apparent (Brien, '70) that the freshwater
sponges are polyphyletic. Recent studies by Racek and Harrison (in preparation) have shown, however, that the various
types of freshwater sponges arose from quite a number of an-
cestors at widely differing times. The clarification of the
position of the non-spongillid freshwater sponges will be a
major problem - involving clarification of their embryology,
ecology, and physiology - inviting further study by sys-
tematists for some time to come.

Particularly exciting discoveries, affecting the taxonomy
of all sponges have occurred within recent years. During the
past 15 years an amazing variety of "living fossil sponges"
have been described, all of which possess relatively massive
calcareous skeletons. These include Sclerospongiae, (Hartman and Goreau, '70), Sphinctozoa (see Hartman, this volume),
and Pharetronida (Vacelet, '70). These discoveries have suggested that the fossil groups, Stromatoporodea and Sphinctozoa, are sponges; furthermore, the Chaetetida have been
transferred from the Phylum Cnidaria to the Porifera (Hart-

man and Goreau, '72). Due to the profound effects which
these discoveries may have on our view of the phylogeny of
the Porifera it is becoming increasingly important that a
"new systematics" be put forth which encompasses both these
new discoveries and previously established fossil groups.
Specifically, a new delineation of higher taxa including
their possible interrelationships is called for.

The systematics of the class Demospongiae continues to
present challenging problems for the establishment of natural
relationships within the group. For example, although Lévi's
subdivision of the group into the subclasses Ceractinomorpha
and Tetractinomorpha (Lévi, '56) has been generally accepted,
Bergquist and Hartman ('69) have concluded, on the basis of
amino acid patterns, that the latter subclass is difficult
to retain in its present context. They further suggest the
abandonment of the order Epipolasida. Numerous problems on
the family and generic level are also outstanding, some of
which are being evaluated through comparative cytology (see
Pomponi, this volume; Simpson, '68b). Further approaches
include serological and transplant techniques for determining
relationships (Connes, et al., '74; Paris, '61) and comparative studies of reproduction (see Chen, this volume, Connes
et al., '74; Lévi, '56) for species delineation.

The basic problem in the taxonomy of the Demospongiae is
the derivation of a set of criteria upon which to base homol-
ogies. Additional, extensive comparative studies are needed
before this can be accomplished.

Life Cycle Events

There are at least three types of biological events which
are cyclic in many sponges; these include sexual reproduction, gemmule formation and hatching, and tissue regression.
The last mentioned involves the loss of the canal system
during winter months and its redevelopment in the spring
(see, for example Simpson, '68a) or the developmentally
similar formation of reduction bodies (Penney, '33; Harrison
et al., '75). Little experimental work or insight into this
phenomenon is available. A second type of cycle which is
apparently universal among sponges is the seasonal production
of gametes. Fell ('74a) has thoroughly reviewed the data on
this subject. Gilbert et al. ('75) and Gilbert ('74) have
recently shown that gamete production in a freshwater species
is probably endogenously controlled; that is, it is independent of environmental stimuli. However, from both these and
other studies (see Fell, '74a and this volume) it is clear
that water temperature can strongly influence the initiation
and/or rate of gametogenesis. Much more experimental work

is required to elucidate the underlying biochemical mechanisms which are responsible for the observed cyclic events. The formation and hatching of gemmules in most freshwater and a few marine species is a third kind of cyclic phenomenon in sponges. The formation of gemmules may also be endogenously controlled (Gilbert, '75) but this situation still requires further investigation. Field data on gemmule formation and hatching has recently been interpreted in terms of an interaction between the environment and the physiological condition of the sponge (Simpson and Fell, '74). Gemmule hatching in the laboratory is strongly affected by temperature but it is not clear if it affects hatching in the field (Simpson and Gilbert, '73). The control of hatching and dormancy may involve changes in cyclic nucleotide metabolism (see Simpson and Rodan, this volume).

Silicon Deposition

We are presently very far from an understanding of the basis of silicon deposition. Recent work on siliceous spicule secretion has demonstrated that a central organic filament (axial filament) and surrounding membrane (silicalemma) are present, probably intracellularly. Silicon is apparently transported by the membrane and polymerized within it (Garrone, '69; Simpson and Vaccaro, '74).

Concentric layering of silicon in spicules has been reported (Schwab and Shore, '71) and is apparently due to differences in water content. However, its significance is not understood. A promising approach to the study of silicon deposition is the use of germanium (Elvin, '72; Simpson and Vaccaro, '74) which, in diatoms (Azam et al., '74), has been shown to be a competitive inhibitor of silicon transport. Since the silicalemma is presumably intracellular, the plasmalemma and other cytoplasmic organelles may also be involved in transport. It has been suggested that the morphology of siliceous spicules may be determined by the morphology of the axial filament (Reiswig, '71), in which case silicon deposition can be viewed as involving a two-component system - the silicalemma which transports silicon and the axial filament which determines the geometry of the polymerized silicon. Germanium apparently uncouples these components producing abnormal bulbous spicules (Elvin, '72; Simpson and Vaccaro, '74; Simpson, unpublished).

Some problems in gemmule physiology

Among the many events which ensue during the hatching of sponge gemmules is the opening of the micropyle. The gemmule coat in freshwater sponges contains much collagen (De-

7

Vos, '72; Harrison and Cowden, '75b) and the opening of the micropyle probably involves the breakdown of the coat in this area. It seems likely that this may involve a collagenase. Investigations of this event have not been reported and are needed in order to gain insight into the means by which the opening of the micropyle is coordinated with the end of germination (cell migration). An even more challenging problem in gemmule physiology concerns the differentiative capacity of the thesocytes. These are apparently a homogenous group of cells able to develop into a fully differentiated young sponge in vitro (Schmidt, '70), which implies the absence of any required morphological pattern of the cells for development. The apparent totipotence of thesocytes, however, has yet to be clearly demonstrated.

Two types of gemmules have been described - diapausing (requiring chilling for hatching) and nondiapausing (not requiring chilling) (Simpson and Fell, '74; Fell, '74b). The biochemical basis of this difference is not known. In the marine sponge Haliclona oculata, it has not yet been possible to germinate the gemmules, which are an example, par excellence, of diapause (Fell, '74b).

The Hexactinellida

This class of sponges is almost unknown on any level other than taxonomic. Their inaccessibility and heavily silicified skeleton account for this. These animals have repeatedly been referred to as substantially synctial (see for example Hyman, '40, Tuzet, '73). However, it is not clear whether this is actually the case or whether the condition of the tissue was altered prior to preservation or fixation. Previous claims of a syncytial pinacoderm in the Demospongiae (Wilson and Penney, '30; Penney, '33) have not been supported by more recent investigations (Bagby, '71; Feige, '69, Harrison, '72; Boury-Esnalt, '73). The whole question of the histological and cytological structure of these animals requires reevaluation using newer methods for in situ fixation.

Distribution of Sponges

The discovery of the fossilized spongillid, Palaeospongilla chubutensis, in lacustrine sediments of the Cretaceous established the date of colonization of freshwater as at least some 100 million years ago. Recent continental drift reconstructions from plate tectonics have allowed the demonstration of feasible pathways of dispersal and speciation leading from this Mesozoic fossil of Patagonia to related extant groups of sponges (Racek and Harrison, '75).

The dispersal of extant sponge species is closely related

to patterns of larval behavior. Environmental factors such as current, turbidity, temperature, light,wave action, and the substratum affect larval dispersal and settling. Intrinsic larval taxic responses to light, gravity, etc., vary from one species to another. Fell ('74) has extensively reviewed all phases of sponge reproduction, including larval behavior and dispersal patterns. A number of marine sponges (see Fell, '74, and in this volume) and freshwater sponges also reproduce asexually through gemmule formation. In these cases, gemmules may be dispersed by currents as plant substrates degenerate (Fell, '74, and in this volume), or they may become detached from shells (Hartman, '58). Gemmule exposure during periods of aridity allows occasional dispersal by wind from dried water beds (Annandale, '15; Fell, '74; Racek and Harrison, '75). Many freshwater sponges are subjected to long intervals of drought interrupted by periods of flooding. These "ephemeral" sponges, found in regions such as the Australian and Sonoran deserts, are capable of completing a compressed life cycle consisting of germination from gemmules, rapid growth, and gemmule formation during the short wet intervals (Harrison, '74a; '75 and discussion in this volume). The ability to survive long periods of desiccation (Harrison, '75) facilitates dispersal by wind in dry periods, and insures species survival until current-mediated gemmule dispersal becomes a factor in periods of flooding. Dispersal of gemmules by aquatic birds has been suggested as a means of explaining the presence of widely separated species populations (Harrison, '71; Racek and Harrison, '75). Although this seems to be a strong possibility, concrete evidence supporting this hypothesis is lacking. There have been no studies of estivating gemmules of the ephemeral species. This would appear to be an exceptionally promising area for biochemical, physiological and ultrastructural study.

Morphogenesis and the Problem of Terminology

Terminology has posed a problem for students of sponge biology. Borojević et al. ('67) have presented a list of definitions and recommended terms, thus bringing some degree of order into the area. However, the problem persists, particularly at the cytological level. This is, in part, due to unanswered questions concerning the morphogenesis and roles of sponge cell types. Central to this issue is the archeocyte, a cell type considered by earlier investigators (Wilson, 1894; Evans, 1899; Minchin, '00) to have blastomeric potential for differentiation.

The morphological and histochemical characteristics of

archeocytes are similar to those of stem cells encountered
in a variety of developmental stages in many organisms, i.e.,
erythroblasts, myoblasts, coelenterate interstitial cells or
neoblasts of turbellarians. Archeocytes have a vesicular
nucleus with a prominent nucleolus and a large extrachromo-
somal volume. They also demonstrate high levels of cytoplasmic RNA (Simpson, '63; Harrison, '74b).

The morphological condition of the nucleus and nucleolus
can frequently be related to synthetic activities of the
cell. The "extended" condition of nuclear chromatin is often
associated with increased RNA transcriptional capacity
(Ringert, '69; Harrison and Cowden, '75a and in this volume).
Conversely, in many instances, a relationship has been established between nuclear chromatin "condensation" and a
decrease in DNA-dependent RNA synthesis (Littau et al., '64).
Although estimates of synthetic activity of cells based
directly upon nucleolar morphology should be approached with
some caution (Barr and Markowitz, '70), archeocytes exhibit
characteristics seen in precursor cells as well as in cells
displaying high levels of synthetic activity. As Simpson
('63) has noted, however, this morphological homogeneity may
or may not be directly related to morphogenetic capacities of
archeocytes. Are all members of this morphological cell
class capable of multifunctional activity, i.e., spicule
formation, collagen synthesis, food transport, excretion,
etc., or is there a division of labor within the class, with
each cell capable of one particular function? Although
Harrison and Cowden ('75a, and in this volume) have demonstrated that deoxyribonucleoprotein organization in larval
and adult archeocytes is comparable, it seems wisest to continue useage of functionally derived nomenclature, i.e.
sclerocyte, fibroblast, etc., once cell-specific substances
are detectable in the course of cytodifferentiation. The
terms "nucleolate" cell or "amebocyte" should be avoided
wherever possible because of ambiguity.

The Nature of Symbiosis in Sponges

A great many freshwater sponges contain intracellular
algae as symbionts. The nature of this symbiosis is not
understood. Recent studies by Gilbert and Allen ('73) have,
however, clarified many aspects of this relationship. These
authors found that Spongilla lacustris is an active, efficient photosynthetic system with a calculated photosynthetic
efficiency of 5.4%. Gross primary productivity values are
linearly related to water temperatures, suggesting that the
stable cellular environment of the host sponge provides
adequate levels of nutrients and a carbon source, thus reduc-

ing limiting factors for photosynthetic rate to temperature and light intensity. Autoradiographic studies following ^{14}C fixation during photosynthesis demonstrated silver grains over algae-free areas of the sponge, suggesting that the photosynthetically-fixed carbon may be excreted from the algae, resulting in transfer to the sponge. Low rates of excretion of photosynthetically-fixed carbon by the sponge (1% of the net primary productivity value) indicate that the sponge is quite efficient in utilizing algae-derived compounds.

Gilbert and Allen ('73) also noted significant seasonal fluctuations in pigment relationships with viable chlorophyll a content dropping significantly immediately prior to the onset of gemmulation, yet rising precipitously at the time of early gemmule initiation (archeocyte aggregation). This intriguing phenomenon is not presently understood and presents an exciting avenue for future studies of the physiological and biochemical interrelationships of sponge life cycle events and symbiosis.

Although the presence of intracellular zoochlorellae definitely aids in survival of sponge species, both in the laboratory and in hotter climates, it is not known if there is a positive selective pressure for acquisition of intracellular zoochlorellae.

Pollution Assay

Because freshwater sponges exist within a well-defined range of physiochemical parameters (Harrison, '74a) it seems likely that at least some species can be utilized as monitors of aquatic environments (Poirrier, in this volume). Recent paleolimnological studies (Racek, '66 and '70; Harrison, unpublished data) indicate that spicular analysis of sediment cores allows one to monitor past environmental fluctuations. This would appear to be a virtually untapped resource for ecological studies - particularly as the species ecology of a great many freshwater sponges is now understood (Harrison, '74a).

Conclusions

As stated earlier, it is impossible to review all aspects of sponge biology within one short chapter. We have not touched upon such areas as the role of sponge collagen in morphogenesis and in the determination of adult morphology, the role of cell surface constituents in reconstitution and cell contact (see Reed et al., in this volume), the possibility of neurohumorally-mediated stimulus transmission and tissue responses in the recognition of foreign substances.

11

We can only emphasize that there are few major animal groups in which so many exciting research problems await investigators.

LITERATURE CITED

Annandale, N. 1915 Fauna of the Chilka Lake: Sponges, Mem. Indian Mus. Calcutta, 5: 23–54.

Arceneaux, Y.A. 1973 Application of starch-gel electrophoresis to the taxonomy of the Spongillidae: Porifera. M.S. Thesis, University of New Orleans, New Orleans, Louisiana, 45 p.

Azam, F., B.B. Hemmingsen and B.E. Volcani 1974 Role of Silicon in diatom metabolism. V. Silicic acid transport and metabolism in the heterotrophic diatom Nitzchia alba. Arch. Microbiol., 97: 103–114.

Bagby, R.M. 1970 The fine structure of pinacocytes in the marine sponge Microciona prolifera. Zeit. Zellforsch., 105: 579–594.

Bakus, G.J. 1965 A refrigerated seawater system for marine organisms. Turtox News, 43: 230–231.

Barr, H.J. and E.H. Markowitz 1970 The development and cytology of Drosophilia embryos genetically deficient for nucleolar organizers. J. Cell Biol., 47: 13–14.

Bergquist, P.R. and W.D. Hartman 1969 Free amino acid patterns and the classification of the Demospongiae. Mar. Biol., 3: 247–268.

Borojević, R. W.G. Fry, W.C. Jones, C. Lévi, R. Rasmont, M. Sarà, and J. Vacelet 1967 Mise au point actuelle de la terminologie des éponges (A reassessment of the terminology for sponges). Bull. Mus. Nat. Hist. Natur., 39: 1224–1235.
"This is the basic review of terminology used in the field of sponge biology."

Boury-Esnalt, N. 1973 L'exopinacoderme des Spongiaires. Bull. Mus. Natr., Paris, 3^{e}ser., n°178, Zoologie, 117: 1193–1206.

Brien, P. 1967a Eponges du Luapula et du lac Moëro. Résult. Scient. Explor. hydrobiol. Bangwelo-Luapula Miss. J.J. Symoens., 11: 1–52.

_____1967b Un nouveau mode de statoblastogénèse chez une éponge d'eau douce africaine: Potamolepis Stendelli (Jaffé) Bull. Acad. r. Belg. Cl. Sci., 53: 552–570.

_____1970 Les potamolepides africaines nouvelles du Luapula et du lac Moëro, In: "Biology of the Porifera, W.G. Fry, ed. Symp. Zool. Soc. Lond., 25: 163–187.
"This paper suggests the establishment of a new family of

freshwater sponges."
Connes, R., J.P. Diaz, G. Nègre, and J. Paris 1974 Etude morphologique, cytologique et serologique de deux formes de Suberites massa de l'etang de Thau. Vie Milieu, 24: 213-224.
DeVos, L. 1972 Fibres géantes de collagens chez l'éponge Ephydatia fluviatilis. Jour. Microscopie, (Paris), 15: 247-252.
Elvin, D.W. 1972 Effect of germanium upon the development of siliceous spicules of some freshwater sponges. Exp. Cell Res., 72: 551-553.
Evans, R. 1899 The structure and metamorphosis of the larva of Spongilla lacustris. Quart. J. Micr. Sci., 43: 363-476.
Feige, W. 1969 Die Feinstruktur der Epitelien von Ephydata fluviatilis. Zool. Jb. Anat., 86: 177-237.
Fell, P.E. 1967 Sponges. In: "Methods in Developmental Biology", F.H. Wilt and N.K. Wessells, eds. Crowell-Collier, New York, pp. 265-276.
"A thorough review of field and laboratory techniques involved in the study of sponges. This work is invaluable to any student of sponge biology."
_____1974a Porifera. In: "Reproduction of Marine Invertebrates, I", A.C. Giese, and J.S. Pearse, eds., Academic Press, New York, pp. 51-132.
"A complete review of the reproductive biology of sponges."
_____1974b Diapause in the gemmules of the marine sponge, Haliclona loosanoffi, with a note on the gemmules of Haliclona occulata. Biol. Bull., 147: 333-351.
Garrone, R. 1969 Collagène, spongine et squelette minéral chez l'éponge Haliclona rosea. Jour. Microscopie (Paris), 8: 581-598.
Gilbert, J.J. 1974 Field experiments on sexuality in the freshwater sponge Spongilla lacustris. The control of oocyte production and the fate of unfertilized oocytes. J. Exp. Zool., 188: 165-178.
_____1975 Field experiments on gemmulation in the freshwater sponge Spongilla lacustris. Trans. Amer. Micros. Soc., 94: 347-356.
Gilbert, J.J. and H.L. Allen 1973 Chlorophyll and primary productivity of some green, freshwater sponges. Int. Rev. ges. Hydrobiol., 58: 633-658.
"One of the few thorough studies of the freshwater sponge-algal symbiosis."
Gilbert, J.J., T.L. Simpson, and G.S. DeNagy 1975 Field experiments on egg production in the fresh-water sponge Spongilla lacustris. Hydrobiologia, 46: 17-27.
Harrison, F.W. 1971 A taxonomical investigation of the

genus Corvomeyemia Weltner (Spongillidae) with an introduction of Corvomeyenia carolinensis sp. nov. Hydrobiologia, 38: 123-140.

_____1972 The nature and role of the basal pinacoderm of Corvomeyenia carolinensis Harrison (Porifera: Spongillidae). A histochemical and developmental study. Hydrobiologia, 39: 495-508.

_____1974a Sponges (Porifera: Spongillidae), in: "Pollution Ecology of Freshwater Invertebrates", C.W. Hart Jr. and S.L.H. Fuller, eds., Academic Press, New York, pp. 29-66. "A species-by-species treatment of the ecology of North American freshwater sponges."

_____1974b Histology and histochemistry of developing outgrowths of Corvomeyenia carolinensis Harrison (Porifera: Spongillidae). J. Morphol., 144: 185-194.

_____1974c The localization of nuclease activity in spongillid gemmules by substrate film enzymology. Acta histochem., 51: 157-163.

Harrison, F.W. and R.R. Cowden 1975a Feulgen microspectrophotometric analysis of deoxyribonucleoprotein organization in larval and adult freshwater sponge nuclei. J. Exp. Zool., 193: 131-136.

_____1975b Cytochemical observations of gemmule development in Eunapius fragilis (Leidy): Porifera; Spongillidae. Differentiation, 4: 99-109.

Harrison, F.W., Dunkelberger and N. Watabe 1975 Cytological examination of reduction bodies of Corvomeyenia carolinensis Harrison (Porifera: Spongillidae). J. Morphol., 145: 483-492.

Hartman, W.D. 1958 Natural history of the marine sponges of southern New England. Bull. Peabody Mus., 12: 1-155.

Hartman, W.D. and T.F. Goreau 1970 Jamaican coralline sponges: Their morphology, ecology, and fossil relatives. In:"Biology of the Porifera". W.G. Fry, ed., Symp. Zool. Soc. Lond., 25: 205-243. "This paper establishes a new class of the phylum Porifera and includes a description of its members."

_____1972 Ceratoporella (Porifera: Sclerospongiae) and the Chaetetid "Corals". Trans. Conn. Acad. Arts Sci., 44: 133-148.

Hyman, L.H. 1940 The Invertebrates. Vol. I. McGraw Hill Book Co. "A classical review of the older literature on sponge biology."

Imlay, M.J. and M.L. Paige 1972 Laboratory growth of freshwater sponges, unionid mussels, and sphaerid clams. Prog. Fish Cult., 34: 210-215.

Lévi, C. 1956 Etude des Halisarca de Roscoff. Embryologie et systematique des Demosponges. Arch. Zool. Exp. Gen., 93: 1-181.
"This monograph establishes two subclasses of the Demospongiae - the Ceractinomorpha and Tetractinomorpha."

Littau, V.C., V.C. Allfrey, J.H. Frenster, and A.E. Mirsky 1964 Active and inactive regions of nuclear chromatin as revealed by electron microscope autoradiography. Proc. Nat. Acad. Sci. U.S.A., 52: 93-100.

Minchin, E.A. 1900 The Porifera and Coelenterata, In: "A Treatise on Zoology". Vol. II. E. Ray Lankester, ed., A. and C. Black, London, pp. 1-188.

Ott, E. and W. Volkheimer 1972 Palaeospongilla chubutensis n.g. et n. sp. - ein Susswasserschwamm aus der Kreide Patagoniens. N. Jb. Geol. Palaont. Abh., 140: 49-63.

Paris, J. 1960 Contribution a la biologie des éponges siliceuses Tethya lyncurium Lmck. et Suberites domuncula O. Histologie des greffes et serologie. Theses. Causse, Grailee, Castelnau, Imprimeur, Montpellier, 74 p.

Penney, J.T. 1933 Reduction and regeneration in fresh-water sponges (Spongilla discoides). J. Exp. Zool., 65: 475-497.

Penney, J.T. and A.A. Racek 1968 Comprehensive revision of a world-wide collection of fresh-water sponges (Porifera: Spongillidae). U.S. Nat. Mus. Bull., 272: 1-184.
"This work is the basis for all present studies in spongillid freshwater sponge systematics."

Poirrier, M.A. 1969 Louisiana fresh-water sponges: Ecology, taxonomy, and distribution. Ph.D. Dissertation, Louisiana State University, Microfilm Inc., Ann Arbor, Michigan, No. 70-9083.
"This, and the following are pioneering studies of the induction of ecomorphic variations and their role in spongillid systematics."

_____1974 Ecomorphic variation in gemmoscleres of Ephydatia fluviatilis Linnaeus (Porifera: Spongillidae) with comments upon its systematics and ecology. Hydrobiologia, 44: 337-347.

Racek, A.A. 1966 Spicular remains of fresh-water sponges. Mem. Conn. Acad. Arts Sci., 17: 78-83.

_____1970 The Porifera, In: "Ianula: An account of the history and development of the Lago di Monterosi, Latium, Italy", G.E. Hutchinson et al., eds. Trans. Amer. Phil. Soc. Philadelphia, n.s., 60: 143-149.
"This study utilizes freshwater sponge spicules in paleolimnology."

Racek, A.A. and F.W. Harrison 1975 The systematic and phylogenetic position of Palaeospongilla chubutensis

(Porifera: Spongillidae). Proc. Linn. Soc. New South Wales, 99: 157-165.
"A study of the evolutionary relationships of the first fully-preserved fossil spongillid to be described."

Reiswig, H.M. 1971 The axial symmetry of sponge spicules and its phylogenetic significance. Cah. biol. mar., 12: 505-514.

Ringertz, N.R. 1969 Cytochemical properties of nuclear proteins and deoxyribonucleoprotein complexes in relation to nuclear function. In: "Handbook of Molecular Cytology", A. Lima-De-Faria, ed., American Elsevier Pub. Co., Inc., New York, pp. 656-684.

Schmidt, I. 1970 Etude preliminaire de la differenciation des thesocytes d'Ephydatia fluviatilis L. extraits mecaniquement de la gemmule. C.R. Acad. Sci., 271: 924-927.

Schwab, D.W. and R.E. Shore 1971 Mechanism of internal stratification of siliceous sponge spicules. Nature, 232: 501-502.

Simpson, T.L. 1963 The biology of the marine sponge Microciona prolifera (Ellis and Solander). I. A study of cellular function and differentiation. J. Exp. Zool., 154: 135-147.

_____1968a The biology of the marine sponge Microciona prolifera (Ellis and Solander). II. Temperature-related annual changes in functional and reproductive elements with a description of larval metamorphosis. Jour. Exp. Mar. Biol. Ecol., 2: 252-277.

_____1968b The structure and function of sponge cells: new criteria for the taxonomy of Poecilosclerid sponges (Demospongiae). Bull. Peabody Mus. Nat. Hist., 25: 1-141.
"This monograph suggests that cytological features are an important and sometimes crucial aspect for establishing taxonomic relationships."

Simpson, T.L. and P.E. Fell 1974 Dormancy among the Porifera: gemmule formation and germination in fresh-water and marine sponges. Trans. Amer. Micros. Soc., 93: 544-577.
"A thorough, critical review of gemmule physiology with emphasis on recent research."

Simpson, T.L. and J.J. Gilbert 1973 Gemmulation, gemmule hatching, and sexual reproduction in fresh-water sponges. I. The life cycle of Spongilla lacustris and Tubella pennsylvanica. Trans. Amer. Mciros. Soc., 92: 422-433.

Simpson, T.L. and C.A. Vaccaro 1974 An ultrastructural study of silica deposition in the fresh-water sponge Spongilla lacustris. Jour. Ultrastruc. Res., 47: 296-309.

Tuzet, O. 1973 Hexactinellides ou Hyalosponges. In: Traite

de Zoologie, III, Fasc. 1, Grasse, P-P, ed.: 633-690.

Vacelet, J. 1970 Les éponges pharetronides actuelle. In: "Biology of the Porifera", W.G. Fry, ed., Symp. Zool. Soc. Lond., 25: 189-204.

Volkheimer, W. and E. Ott 1973 Esponges de aqua dulce del Cretacio de la Patagonia. Nuevos datos acerca de su posicon sistematica y su importancia paleobiogeografica y paleoclimatologica. Act. Quinto Cong. Geol. Argentino, 3: 455-461.

Wilson, H.V. 1894 Observations on the gemmule and egg development in marine sponges. J. Morphol., 9: 277-406.

Wilson, H.V. and J.T. Penney 1930 The regeneration of sponges (Microciona) from dissociated cells. J. Exp. Zool., 56: 73-147.

General Discussion:
Problems in Sponge Biology

Topic 1. Does Asexual Production of Sponge Larvae Occur?

HARRISON: As he has written the most recent review article which relates to the first topic, I'll ask Paul Fell to introduce this topic.

FELL: Since Wilson first described asexual reproduction of larvae, there has been a debate concerning whether this, in fact, occurs. I think that, if you look at the literature, it seems probable that this occurs in some cases. However, in many cases there is uncertainty about the details of reproduction or asexual reproduction. I think that one shouldn't jump too easily to the conclusion that larvae are being produced asexually. There are cases, for instance, where sperm are present for only a very short period of time, when the details of oocyte growth are unclear, and/or when details of early embryonic development are unclear. Given these circumstances, one could be led to the conclusion that the larvae are asexually produced when, in fact, with a more detailed study, a sexual reproductive sequence might be demonstrated. I think we need to have more studies of the reproduction of sponges and to have studies that are more detailed. Maybe we should suggest some criteria we should use in trying to determine whether the normal method of reproduction in a given species is by the asexual or sexual process or, as has been suggested in some cases, by both methods.

HARRISON: What do you think of Dr. Patricia Bergquist's idea of the advantages of asexual reproduction in the intertidal environment? Perhaps we should concentrate our efforts toward studies of intertidal species in an attempt to determine if this really is a valid reproductive force.

FELL: I fail to see the real advantages to an intertidal situation that she suggests, especially since many intertidal sponges are demonstrated to produce larvae sexually. Also, in some of the species that she described, Halichondria panicea and some of the haliclonids, sexual processes have been demonstrated in either the same species, subspecies, or related species. I don't see a clear advantage to asexual reproduction in the intertidal, as she has suggested it. That doesn't mean that it doesn't exist under the particular circumstances that occur there.

RUDE: I talked to Dr. Bergquist several months ago. In one of the species which she had previously believed to be reproducing only asexually, Halichondria moorei, she and her co-workers have now found at least one case of sexual reproduction. I don't know but that she might have modified her ideas on asexual reproduction of larvae.

REISWIG: We all know that sexual and asexual reproductive processes happen in a number of phyla other than sponges. However, the concept of undifferentiated archeocytes wandering together to form a somatic cell aggregation and from this aggregation forming a very complex, very tightly controlled, behaviorally complex larval stage identical to that produced sexually is difficult to accept. I can accept the idea of production of some kind of asexual reproductive structure by this process, perhaps a bud, but I cannot see how an aggregation of undifferentiated amoebocytes can form such a complex integrated system as a parenchymula larva, which is not only structurally complex but behaviorally complex. These are concepts that make it difficult for many of us to accept the idea of asexual production of larvae.

FELL: In a number of animals, aggregates of the type you describe have been shown to occur. In planarians, for example, the blastomeres are separated within a yolk mass, and then they aggregate to form an embryo. In certain of the annual fishes, dispersed blastomeres come together to form an embryonic shield. So, that this in itself doesn't bother me too much, but I think there is a real problem in interpreting the developmental stages in embryos. One thing that leads to confusion, too, is that the aggregation of nurse cells does occur in many cases. Especially where cytological details are not clear because of the abundance of reserve granules and other cytoplasmic inclusions, one might believe that these aggregating cells are coming together to form an embryo asexually when, in fact, one is looking at oogenesis. Even in sponges that have been studied in some detail, like Microciona, it is hard to hypothesize an orderly sequence of events in later oogenesis and early embryonic development. In many of these cases, it is hard to know what one is dealing with. In some cases, people have described nurse cells surrounding embryos and suggested that, once a cleavage has begun, there is growth of the embryo by addition of cells or cell parts.

SIMPSON: Maybe we should look at one case that has been studied in some detail, the freshwater species Spongilla

lacustris. Although the fine cytological details of fertilization and all the specific details leading up to the production of parenchymula larvae have not been seen at every
life-cycle stage of this species, the sequence of production
of gametes and formation of larvae occurs annually in a very
regular pattern. Dr. John Gilbert has established that, if
gemmules are transferred back to the field later in the sum-
mer--it so happens that all those transplanted were female--
when there is no available sperm for fertilization, those
eggs produced by the gemmule-derived sponges actually undergo
degeneration. In S. lacustris, even though it is not an
oviparous animal, the production of parenchymula larvae is a
sexual process. I wanted to make one other statement related
to Henry Reiswig's comment. The problem that both he and I
have arises from looking at adult mesenchymal cells aggregating. We are discussing a differentiated sexual product,
which becomes mesenchymal, then without fertilization aggregates to form a morphological structure, a larva, which we
usually think of as only a sexual product. The concept of
asexual reproduction of larvae is a difficult one because
the organism arrives at exactly the same stage that it does
in sexual reproduction. In other organisms this, in general,
is not true, or at least you can follow the pathways which
lead to asexual versus sexual production of a stage. In this
case we have to accept that the larva is a highly differentiated, ciliated, phototactic structure that has an area
specialized for attaching to the substratum and that this
larva can arise in an identical fashion, either without fertilization, through aggregation of adult cells, or by fertilization. It is more a matter of accepting the conceptual
problem than of anything else, because it would be quite important to document that adult mesenchymal cells aggregate
and differentiate in exactly the same way as the fertilized
egg does.

HARRISON: Dr. Ronald Cowden and I have found that larval
and somatic archeocytes are almost identical in terms of
deoxyribonucleoprotein organization. When one talks about
a larval archeocyte and an archeocyte found in an adult
sponge, one is essentially talking about the same cell type
in terms of organization of the deoxyribonucleoprotein complex. I'm not sure that these archeocytes are at all different. Archeocytes in larvae and archeocytes of adult
sponges may have the same developmental potential, and, as
such, an aggregate of somatic archeocytes could have the
capability of forming a larva.

SIMPSON: It doesn't surprise me that a mature parenchymula larvae would have the same characteristics as the adult tissue. If you look at enough mature parenchymula larvae in a number of sponges, you get the decided feeling that potentially everything is there for the adult. The larva is in a modified free-swimming state, but the larva shows cellular differentiation, not a mass of relatively undifferentiated cells. It is at least comparable to a late gastrula.

HARRISON: In our cytochemical study of larval development, Dr. Cowden and I considered the escape larva to be a miniature adult. I think that this matter of asexual production of larvae is something we need to investigate. If a collaborative effort is required to solve this problem, I'm willing to cooperate.

Topic 2. Intraspecific Strain Specificity--Its Reality, Its Genetic Interpretation And Its Implications Vis-a-vis Speciation.

HARRISON: This topic refers to the work done with Ephydatia fluviatilis by Dr. R. Rasmont's group. Sponges of the same species, but collected from different ponds, will remain strain specific. If reared side by side, they will not merge, but will remain as distinct entities. In fact, alpha, beta, and gamma strains have separate and distinct growth patterns.

REISWIG: This phenomenon, intraspecific strain specificity, which occurs within populations, may not have taxonomic validity. It could be simply a process ensuring out-breeding and sexual crossing. If sperm are produced by a normal sponge individual, and if one expects out-cross breeding to be selectively advantageous, one would expect this individual to recognize and reject its own sperm. It may be that this intraspecific strain specificity, this ability of sponges to recognize other specimens within its population, could be the mechanism of simply rejecting its own sperm and recognizing sperm from another strain. The sponge would also recognize sperm from other sponges of the same strain. This would ensure fertilization by another element of the same species population. In one way this could be a selective advantage to what appears to be a general condition of sponge populations. It is now known that this situation

holds for most marine sponge populations, as well as freshwater sponges.

Topic 3.   The Fate of Nurse Cells and Their
           Genetic Complements

FELL: As I mentioned briefly before, there have been descriptions not only of nurse cells around oocytes but also
of nurse cells associated with embryos. I think in some of
these cases it is again a matter of interpretation. In some
cases, people have described as mature oocytes what are actually small oocytes and have confused the larger oocytes with
embryos. In some cases, therefore, where nurse cells have
been described around embryos, this may in fact be nurse
cells around oocytes. In other sponges, especially in the
Calcarea, it appears that nurse cells are associated with
larvae. There has been a suggestion that these exist even
in a relatively late stage of the life history. In these
cases, the question could be asked whether these cells actually persist as somatic cells within the new individual or
whether they are phagocytized and used as nutrient material.
In most cases where nurse cells are definitely associated
with oocytes, the cells become phagocytized. In some cases,
there are quite elaborate yolk spheres produced. I think
that, in this case, the answer to the question whether the
genetic components persist is probably that they don't. In
other cases, as in some of the haliclonids, the nurse cells
remain intact in the oocyte and even within the early embryos, although nuclei disappear very soon after the cells
are engulfed by the oocyte. We don't know what happens to
the nucleus or the nuclear components in these cases. At
least the cytoplasm persists until a relatively late stage.
Even at the electron microscopic level, the cytoplasm of
these cells seems not to undergo any dramatic change, at
least until after embryogenesis has been initiated.

CHEN: I have seen a case in which maternal cells of different histological types serve as nurse cells for oocytes
and embryos. In Halisarca nahantensis, there is an increasing number of "nucleolate" cells surrounding the oocytes at
the vitellogenic phase. The oocytes extend their lobe-like
cytoplasmic processes to contact the filiform processes of
"nucleolate" cells. After vitellogenesis is complete and
the oocyte is enclosed in the endopinacocyte-lined cyst,

these "nucleolate" cells no longer surround the oocyte. On
the other hand, the maternal fuchsinophil cells increase in
number and surround the embryo at the late blastula stage.
The fuchsinophil cells migrate from the maternal mesohyl into
the nurse cavity and, finally, the larva takes in fuchsinophil cells through an opening among the posterior larval
cells. Thus, the embryos increase in size during larval
formation. Ultrastructural studies on the larva show that
the fuchsinophil cells are being degraded. This evidence has
led me to suggest that "nucleolate" cells are involved in
vitellogenesis of the oocyte and fuchsinophil cells are associated with embryonic development. Both cell types might
be considered as nurse cells.

Topic 4. Is The "Carrier-Cell" System Of Fertilization In
Sponges Supported By Convincing Evidence?

REISWIG: One of Dr. Fell's early remarks in his recent
paper on sponge reproduction concerns the problem of drawing
conclusions about dynamic events from looking at static sections in fixed material. Especially related to this are
questions of the morphogenetic pathways of choanocytes and
archeocytes. All the information we have about fertilization in sponges and the mechanism of sperm transfer unfortunately comes from exactly this type of material. It involves brief glimpses of material fixed at a point in time
and hypotheses about the dynamic events of fertilization
from these observations. We know that egg cells and/or
early oocytes are fed particulate materials by nurse cells
or trophocytes surrounding them. In many cases, the published figures of apparent feeding stages are identical to those
figures that have been interpreted as sperm fertilization
events. In fact, in the year following his first description
of fertilization, Dr.-J. Gatenby made a qualification of his
earlier observation. Apparently, several of his published
figures of fertilization are , in fact, egg feeding. So
many of the figures of fertilization in Grantia that have
been reproduced over and over again, in Hyman's "The Invertebrates", Volume I, and subsequent books, are open to seri-
ous question about whether this is fertilization or egg
feeding. As far as I know, no one has really attacked the
problem in an experimental way by giving a virginal egg-
bearing female sponge known sperm at a given time and following the process of fertilization sequentially. This needs

to be done if we want to know how penetration of the egg is accomplished.

CHEN: If we obtain a case where sperm-release occurs in a short period of time but oogenesis occurs in a longer period, histological information might indicate something about fertilization in sponges. In Halisarca nahantensis, sperm-release occurs in April while oogenesis occurs from November through May. Many dark acidophilic masses were found to surround a group of choanocytes during the period of sperm-release. These darkly stained masses, containing 5 to 8 nuclei of choanocytes (choanocyte cytoplasm could not be distinguished), move to the mesohyl and attach to the oocytes undergoing vitellogenesis. Some remnants of these masses were found to attach to the follicles of mature oocytes. Later, two pronuclei appeared in the mature oocyte. Because of the strong stainability of these masses, it is difficult to find if there are any sperm inside them. However, it is possible that sperm of Halisarca might stimulate differentiation of a group of choanocytes which will serve as carriers to transport sperm to the oocytes.

REISWIG: Did you control the fertilization in these animals, or were sperm engulfed or caught at a given time before you cut the sections?

CHEN: I have only followed reproduction of Halisarca by weekly collection from their natural habitat. One of the species I have studied, H. dujardini, is dioecious and its gametogenesis is synchronous. In the other species, H. nahantensis, oogenesis is also synchronous. Both species breed annually and reproductive events repeat at the same period of each year. Information on the timing and mode of reproduction and development of the oocytes will be the basis of understanding fertilization in sponges.

FELL: The type of system you described is the kind that really needs to be utilized because, in most sponges having internal development, the developmental sequence is asynchronous. In trying to do the type experiment that Dr. Reiswig suggested, there would be great technical difficulties in trying to find the few oocytes that are in the process of fertilization at a particular time.

SIMPSON: With finding that a particular northern population of Spongilla lacustris is dioecious it is conceivable, because it is known when the eggs are produced and when the sperm is released, to collect a sponge and section it the same day, find out if it is a female, tag it, go back and find a male, and to artificially fertilize the female sponge in the field. We can then sample since the timing of gametogenesis is known and thus, I think, one has a greater chance of synchrony. But, even in that case, there may be different areas of the same sponge with different rates of penetration and transport of the sperm cells.

There is a real problem about how the sperm cells get to the eggs. I wounder if they have to be transported in some way? This could be inside cells or not, or possibly the transporting cells have to push them along. Even if the distances are quite small, 5 or 10 microns, it is a long way to have to move in ground substance.

CHEN: Another way to approach this is to look at the morphology of the sperm. In Halisarca, the morphology of the sperm is quite different from that of other invertebrates and sponges yet described. The sperm move with a quick, slight jerk and with the head turned backward, so that the point of attachment of the flagellum is foremost. I cannot demonstrate any structure of the acrosome at the light microscopic level. Complete condensation of nuclear material and elimination of cytoplasm are not observed in the sperm released from the sponge. I think these morphological features should be examined at the ultrastructural level in order to determine how the sperm function in the fertilization mechanism.

SIMPSON: The experiment with populations of dioecious freshwater sponges could be done in an elegant way because very young male animals could be collected and identified immediately. These could then be exposed to tritiated thymidine, and then the tritiated products could be traced. That would require a lot of work, but it would give some clear answers about the pathway of sperm transport to the eggs. The answer to the question of specific involvement of cell membranes and structures then would require electron microscopy. I think it would be unlikely that fertilization mechanisms share no relationship to each other in marine and freshwater sponges. It is possible that some species of marine sponges are very highly specialized in the method of sperm transport, and specialized mechanisms may not have evolved in the freshwater sponges. Any sequence which can be established is one more than we have now.

Topic 5.   Color: What Is The Significance Of Coloration
In Sponges?

LITCHFIELD: Looking at this topic from a biochemist's view-
point, I can think of three possible reasons for colors in
sponges.   The first possibility involves the carotenoid pigments found in most sponges.   It is well documented that in
certain microorganisms carotenoids serve as photoprotectors.
This means that excessive solar radiation will preferential-
ly destroy the carotenoid pigments before it will decompose
vital metabolic products essential to the animal.   Secondly,
color in sponges could be a matter of recognition.   We know
that certain sponges are toxic or indigestible by their
color.   Finally, coloration could be fortuitous, simply a
metabolic waste product.   The bright colors of maple leaves
in the fall, for example, serve no biochemical function but
are the product of a decomposition process going on within
the leaf.   Paul Fell mentioned that he has observed examples
of Haliclona exhibiting more intense coloration on the upper
surfaces than on the underside.   This pattern could be attributed to carotenoids.

FELL: Yes, where sponges were attached to the side of a
piling or similar structure, the part exposed to the sun
might be pigmented and the part below would not be.   If you
looked at different specimens, those in the sun would be pigmented and those in the shaded areas would not be.

LITCHFIELD: This would correlate with the photoprotection
hypothesis, although obviously would not prove it.

LITTLE: You find this, also, in the coral reef sponges in
the western Gulf of Mexico.

WIEDENMAYER: It might be suggested that the function of
color in sponges can vary from taxon to taxon.   The keratose
sponge "Verongia" fistularis is generally yellow in life,
but turns dark blue, then black, when removed from water.
Some West Indian species, in the same and in different
keratose genera, with a similar necrotic change, also have
a yellow choanosome in life, but display a remarkable poly-
chromy of the surface.   These are notably "Verongia" cauli-
formis, Verongula gigantea, "Aplysina" fenestrata and
"Ianthella" ianthella.   The latter is most astonishing: I
have seen drab, red-brown, golden yellow, orange, salmon,
pink, carmine red, and deep purple specimens, all in the same

habitat and general locality. The West Indian haplosclerid "Callyspongia" plicifera varies from pink to purple, orange, and blue. Its color in life is made even more remarkable by its iridescence or luminescence, a property which is absent in other West Indian sponges of this genus. Haliclona aquae-ductus is also strikingly polychromous. The color variants reported by Italian authors for populations of the Mediterranean (greyish and yellowish white, drab, pink, brownish orange) complement those which I observed in the Bahamas (pinkish lavender to purple, drab, olive, greyish). Other species of Haliclona in the West Indies seem to be stable in color. Yet, such phenomena have not been investigated properly.

HARRISON: Dr. Resh, in your work with predators on sponges, did you notice any relationship between predation and color preferences?

RESH: No. There seemed to be situations involving patches of sponge where one would expect predators and find none, neither spongilla flies nor caddis flies.

POIRRIER: I agree. I've worked with spongilla flies in freshwater sponges. Some of the freshwater sponges in acidic, highly colored swamp habitats can be of a very dark black color. There doesn't appear to be any absence of predators or parasites, such as spongilla flies, among these.

HARRISON: So, in terms of predation, you saw no relationship?

MORALES: It seems that the central problem here is, if one has an organism that is feeding on a sponge, the organism has obviously become adapted to feeding on it and will not shy away from the sponge even when there is a color difference.

POIRRIER: Well, I certainly wouldn't conclude that there is no correlation between predation and coloration in sponges. I'm just simply saying that, in the situation I've observed with spongilla flies and a few species of freshwater sponges, there was no apparent relationship.

DAYTON: In our Antarctic community, the predators on sponges are starfish and are quite blind. However, the brightest sponges are the ones that are never eaten. With the exception of the white ones, this shows a perfect correlation.

We have several brightly colored sponges such as <u>Latrunculia</u> <u>apicalis</u> which is green, <u>Dendrilla</u> <u>membranosa</u> and <u>Isodictya</u> <u>erinacea</u> which are yellow and <u>Leucetta</u> <u>leptorhapsis</u> which is a conspicuous white. These species have little or no spicule protection, but are never eaten. But, starfish are blind, so I don't know that there is a visual warning in this case.

HARTMAN: Are starfish, in fact, really blind? What about the pigmented podia at the ends of the arms? In some starfish, at least, these terminal sensory podia bear photoreceptors that might be able to differentiate a white sponge.

DAYTON: I'd be delighted if someone would demonstrate that.

HARRISON: I was thinking more of a chemoreceptor system in predators. I wonder if the correlation isn't with color, but with a diffusible product related to the color. In other words, the pigment itself may be a noxious product.

HARTMAN: There is the very interesting fact that when zoanthideans live on sponges they are often colored in a complimentary way. Under these circumstances the cnidarian symbionts must be quite obvious to predators that might thus be deterred from eating the sponge. In studies of fish feeding on sponges, Randall and I looked fairly closely at whether there might be a correlation with color, that perhaps some of the more brightly colored sponges were in fact exhibiting warning colors that result in avoidance by fish. We were unable to find any evidence of this in the fish feeding on sponges in the West Indies. Most of the sponge-feeding fishes seem to be generalists with respect to their prey. They eat little pieces of many types of sponges and they don't seem to specialize in any particular color or show avoidance of any color, which is surprising. There is also the fact that in deep water, between 300-400 feet where light is reduced, the sponges, if anything, become brighter in color. When one goes down along the vertical wall of the deep fore-reef in the Caribbean in a submarine with lights on, one is impressed with the extraordinary colors of the sponges existing there. Yet to a fish or any other predator these must appear to be various shades of gray. It could just be that a red sponge looks unusually black, and that this color could have some protective value. The few observations that Randall and I made, however, don't really provide enough information to say that the color of sponges is significant in terms of predator avoidance.

RUDE: One case supporting the idea that color might be a protection against light damage to some metabolites in the sponge is seen in one of the areas off San Diego where I'm working. Here there is a sponge, Stellata sp., which appears to be polymorphic in color. There is a pure white strain and a slate gray strain which I cannot distinguish according to spicules or the tissue. They occur adjacent to each other with the exception that the white form occurs only in shaded habitats whereas the gray form occurs in both shaded habitats and in those exposed to light. If the gray coloration is a protection against some sort of light damage, that might explain the distribution.

HARTMAN: Certainly there are many examples, many of them detailed by Sarà, where in caves in the Mediterranean sponge color becomes lighter as one goes deeper into caves.

RUETZLER: Yes, for example Chondrilla nucula and Petrosia ficiformis where all transitions from dark brown to brown or purplish brown, pink and white, occur with a gradient of decreasing light. Related to an earlier point is the fact that a predator of Petrosia ficiformis, the opisthobranch Peltodoris atromaculata adjusts its color pattern to that of the sponge on which it lives. Presumably, this is a strategy of camouflage as it is not uncommon for this kind of association. I have observed crabs, mites and even kinorhynchs on bright red axinellids in Tunisia which closely imitate the color of their hosts.

FULLER: I'd like to make a point and ask a question. Color vision--perhaps that's too broad a description of the phenomena--some version of color vision as we think of it is known in fishes. The question is this: is protective mimicry known among sponges, particularly highly colored or highly sophisticated morphological forms of marine sponges?

RUETZLER: For that I have no evidence. On the other hand, could lack of color vision in some predators or symbionts not partly be replaced by temperature sensing since the various pigments would absorb different portions of the available radiant energy?

FULLER: I'm sure that's a very valid possibility.

RESH: I would like to agree with you on the point of predators' adopting characteristics of the sponge. The caddis flies, predators on freshwater sponges, contain a genus with both sponge-feeding and non-feeding species. Characteristically, as caddis flies build cases, they will usually use sand particles. Certain species of sponge feeders will actually incorporate host sponge into their cases, and very often they are indistinguishable from the sponge.

LITCHFIELD: One should also recognize the human bias in talking about sponge coloration and possible functions. Since we are human beings, we think mainly of the range of colors that we can see. There are two other factors. Certainly the color of sponges changes as you go deeper in the ocean. In addition, the color seen depends on the wavelength sensitivity of the organism looking at a sponge. Certain insects, for example, can see longwave ultraviolet radiation that humans cannot see.

DAYTON: Are there any of these pigments which might also taste bad? We seem to assume that these colors are predator defenses and yet, whenever anybody asks for evidence that they are, there is none. The issue of protective mimicry, adaptive coloration, etc., all involve the notion of the brightly colored sponges' tasting bad. Maybe there are some mimics, but we don't know what the model is, much less, whether it is distasteful. The fish eat the brightly colored ones. The brightly colored sponges seem to show up down deeper. It is almost as though there were a negative selection of bright colors. Up where the colors show, they might be eaten more, and there still doesn't seem to be much reason to suspect that the colors serve as a predator defense.

Topic 6. Factors Affecting Gemmule Formation and Hatching

POIRRIER: In working with Ephydatia fluviatilis gemmules which were recently formed in the laboratory, I accidentally lowered the pH to below 4; of course, it rose again, but these gemmules hatched within the next couple of days. We repeated this with other species and got the same response. I don't know how this fits into your scheme of thought in terms of factors that may control gemmule hatching, but, nevertheless, this is an experimental observation.

SIMPSON: Is there any correlation in your field data?

POIRRIER: We've gotten pH values as low as pH 3.9 in some of these sponge habitats, so I imagine the same thing could occur in nature. In other words, you might get the same response at a higher pH value just by pH fluctuation in the field. I worked also with an alga, <u>Pithophoria</u>, which has a substance which inhibits the germination of the akinete, a resistant structure. This mechanism works only under alkaline conditions. There may be something like that occurring in the freshwater sponges.

SIMPSON: Particularly in some of the habitats that you're looking at, are pH cycles known?

POIRRIER: Yes. There could be seasonal cycles. When these habitats fill in the spring they have very low pH values, whereas in the fall during low water conditions they have high pH values.

SIMPSON: Does that generally correlate with gemmule formation and hatching?

POIRRIER: No. The freshwater sponges which I have investigated have definite annual cycles in some habitats. When they reach a certain age and size, they produce gemmules. One might find a different cycle in another habitat or in the presence of different weather conditions. Sudden onset of cold weather or a hard rain which stirs the bottom might disrupt the normal cycle, making the cycle askew from that of the normal population. In other words gemmule production appears to be habitat-dependent in many species, as opposed to something inherent in the genetics of the species.

SIMPSON: But in general are you saying that the sponge tends to override any asynchrony in cycles from one habitat to the next?

POIRRIER: I could point to what I call normal cycles in certain habitats, but then again I could point to habitats where the sponge does not follow these cycles. We have some flowing wells with constant temperatures where sponges are found which do not normally remain active during the winter in Louisiana.

SIMPSON: You are impressed by the fact that in the generalized picture there is some normalcy which, if you move away from, there can be some override of the response?

POIRRIER: Right. Exactly.

HARRISON: May I interject something here? There is one
major ecological group of freshwater sponges which I consider
the ephemerals. Examples are found in the Sonoran Desert and
in the heart of Australia. The gemmules of these forms remain viable in nature for long periods of time. There are
also cases in which gemmules of these species have germinated
after having been dry on the museum shelf for 25 years. If
they are placed into water, they germinate. In one species,
<u>Heterorotula</u> <u>multidentata</u>, from the Finke River of Australia,
not only was there a history of gemmules remaining viable
over a six year drought, but also the gemmule coats retained
nuclease activity. In other words, freshwater sponge gemmule
coats are more than a protective coat of collagen and matrix.
The gemmule coat is enzymatically active with ribonuclease
and deoxyribonuclease activity present.

DE NAGY: In gemmule formation in the ephemerals, was des-
sication of the living sponge the key for gemmule formation,
or was it necessary for the gemmule to form in order for the
sponges to survive initial lowering of the water level?

HARRISON: I don't know. In the areas where these species
occur, it is a matter of finding sponge gemmules in a dry
stream bed and checking with the weather bureau to determine
the last rainfall in the vicinity.

POIRRIER: Many habitats in Louisiana are temporary habitats,
and we have freshwater sponges living in them. These habitats are normally dry throughout the summer months. I have
photographs of freshwater sponges in trees, for instance, in
Catahoula Lake in Louisiana which is forty feet deep during
the winter. In the summer you can drive around it in a jeep
and collect sponges from the trees. Much of my summer collecting is from dry roadside ditches which are overgrown with
vegetation. One can walk along and scrape the freshwater
sponge gemmules off the button bushes. These gemmules are
highly resistant to desiccation. I've had students take
gemmules out of my dry collection, now at least 10 years old,
and obtain germination by placing them in proper water.

LITTLE: I think I have a slide which shows sponges in the
tops of trees in the Amazon basin.

POIRRIER: This gets into the business of ecomorphic variation, also, because there appears to be a response in the

structure of the gemmules related to whether or not the habitat will become dry. In some habitats which never dry, the gemmoscleres are not as robust. One doesn't find a heavy pneumatic layer on the gemmule, whereas in those habitats which do dry there is a thick pneumatic layer. The sponge appears to be adapted to drying, and the key is the water quality of the habitat. As the habitat conditions become more alkaline there is this response. This is the key that tells the sponge that the habitat is drying.

SIMPSON: You have that correlation?

POIRRIER: I now have with a number of different species, and it makes sense.

LITTLE: In constant flowing areas of fairly uniform water is there a temperature factor?

POIRRIER: I don't know if I can say that. There may be.

LITTLE: I have noticed this in some aquaria where I have raised the temperature one degree Centigrade. In two days the sponge is gone, and there are gemmules present.

POIRRIER: In terms of the response of gemmule formation, this is right. One of the questions I would like to throw out is this: does anyone know if photoperiod is involved? In other words, has anyone done any experimental work that may indicate that gemmule formation and hatching are in any way triggered by photoperiod?

SIMPSON: Dr. R. Rasmont published a short paper in 1970 in which he describes some experiments in which he was able to inhibit gemmule formation in constant illumination. These sponges contained no, or at least very few, symbionts and it was suggested that the inhibition was due to a direct photochemical effect on the sponges. I want to make one additional comment about your observations of pH, which I hadn't previously heard about. It has been demonstrated in fibroblast cultures that changes of pH are capable of affecting the cyclic nucleotide levels. We've found cyclic nucleotides are likely a central feature of the germination of gemmules and probably gemmule formation, as well.

POIRRIER: I mentioned the pH change more as an experimental observation, but, again, I don't think this is _the_ mechanism that is occurring, although it may be related to it.

SIMPSON: But you say you do have a correlation between the increase in alkalinity and the formation of gemmules?

POIRRIER: Yes. I have this correlation with the morphology of the gemmule, also. I noticed in one of your papers that you referred to a variety of Spongilla lacustris without the pneumatic layer present in the gemmule coat. I can induce the formation of gemmoscleres and the gemmule coat pneumatic layer in Spongilla lacustris by modifying the physiocochemical characteristics of the water. I can also modify the morphology of the gemmoscleres.

SIMPSON: Is the cause primarily pH?

POIRRIER: I don't know what it is. It's interrelated, I hate to pin it down, but, anyhow, I can do it experimentally.

SIMPSON: It would be very interesting to look at the temperature readings and climatic changes which may lead to vegetation decomposition and thus to changes in pH.

POIRRIER: But it would be nice to have experimental data which would indicate the minimal shift in pH initiating a response in terms of gemmule hatching. Slight shifts, I think, would be important.

DE NAGY: Dr. Poirrier talked about sponges' not gemmulating in a constant environment. I wondered if there are any sponges that are unable to gemmulate when living in a constant environment that cannot be induced to gemmulate when shifted to a more stressful and/or more seasonal environment?

POIRRIER: Well, let me venture one thing. All the species that I've worked with in Louisiana and other Southern states, if brought into the lab in an immature stage and put into a covered stacking dish in habitat water, will usually form gemmules within two weeks. It doesn't make any difference if you put them in a refrigerator or in darkness. They will form gemmules in two weeks. I don't know whether this is due to starvation or what, but they do this. I've used this in order to identify sponges without gemmules and to work with ecomorphic variation.

SIMPSON: Do you know what the pH of your water is?

POIRRIER: I have never really monitored it. I have just

used this as a mechanism to identify immature sponges without gemmules. I have done this with all of those I have collected without gemmules from a habitat.

SIMPSON: You always use habitat water?

POIRRIER: Always habitat water. I have allowed it to evaporate a little bit. One can get ecomorphic variation if it evaporates very much.

FELL: So far the discussion has been on freshwater sponges, but there are some marine sponges that produce gemmules. One of the factors that may be important in the regulation of germination in brackish-water species is salinity. Within a certain salinity range, germination will occur. Outside that range you get reversible inhibition of germination up to a certain point. We found that the gemmules of Haliclona loosanoffi from Connecticut will germinate within a range of 20 to 40 parts per thousand. No germination occurs outside that range, but you get reversible inhibition of germination at 5, 10, 15 and 45 parts per thousand. It appears that the gemmules don't survive exposure to fresh water, at least a long exposure. With gemmules from a habitat where the usual salinity range is from about 20 to 30 parts per thousand, this is the situation. Haliclona loosanoffi is also found in habitats where the usual salinity is considerably lower than this. I would think that in these situations the gemmules would have a lower range in which they germinate and probably a lower tolerance, and also, could possibly tolerate exposure to fresh water.

CHEN: There is one species of Haliclona which exists subtidally in a constant environment. From April through August every year, gemmules constantly appear inside the sponge.

FELL: The occurrence of gemmules is seasonal even though the environment is so uniform?

CHEN: Yes.

SIMPSON; I'd like to suggest that those of us who are working on gemmules get together--particularly with those who are working in the South. There are some very interesting experiments that could be worked out quite simply by transporting gemmules from the South to the North, and the reverse, and then following the cycling. This problem of an

endogenous rhythmic cycle that the animals may have, which
may or may not be overridden by environment, may also be determined by the genetics of the species. This is at least
one very excellent way to get at it. In fact, I think it
would be quite easy to do if we could convince the airlines
to ship the material.

HARRISON: There is one other possibility, too. We could
get shipments of gemmules out of Australia, also.

REISWIG: Has anyone tried any exposure of gemmules to liquid
nitrogen for purposes of transport?

SIMPSON: No, but we lyophilized some last summer. These
gemmules had been stored for almost nine months,but, after
about eight months of storage germination becomes more and
more difficult indicating that there are probably endogenous
events occurring inside the gemmule, at least in Spongilla
lacustris. I was encouraged by what we found. We could get
germination after lyophilization if we placed the gemmules
in 0.1 N sodium chloride for half an hour, brought them
down to intermediate concentration, and then brought them
down to pond water. We made one very interesting observation during this study. When the sponges hatched out, they
no longer were green. These were sponges from gemmules that
had been originally filled with symbiotic algae and the
sponges were now completely white. This may be a mechanism
for producing symbiontless sponges. I think that it's very
encouraging. I also think that, if the gemmules were lyophilized in January after they were stored for only a month
under refrigeration, the results would be much better. This
needs to be tried and would simplify the problem of mailing
gemmules world-wide whenever one is looking at the question
of endogenous rhythms.

RESH: There are habitat-dependent factors, too. Spongilla
lacustris will gemmulate in the fall, but the colonial form
exists all year. Has anyone looked at this, for instance,
in the North? I know that in Kentucky and Indiana, we find
overwintering of sponges entirely as gemmules. Has anyone
found the definitive colonial sponge occurring in the winter
months during any of the studies he has done in the North?
Is there a point where you find this relationship occurring?

HARRISON: One species with which I was working survived best
in South Carolina during the winter months. The colonial
form was often eliminated by heat. Gemmulation was stimu-

38

lated in this species by the onset of summer.

POIRRIER: This is the typical life history of Ephydatia
fluviatilis in Louisiana. The colonial form occurs only during winter. As soon as the water temperature gets up to
30°C, gemmules form. With the first cool weather in September, germination occurs, and young colonies develop. The
colonial sponge is completely absent during the summer!
Other species behave like this (Anheteromeyenia ryderi, for
instance).

SIMPSON: That is really amazing! Those are the species
that would really be interesting to look at if they were
transplanted to different latitudes.

FELL: One has the same situation in Haliclona loosanoffi, a
marine sponge. In North Carolina it oversummers in the form
of gemmules, in New England it overwinters in the form of
gemmules, and in Virginia colonial specimens occur throughout the entire year. In North Carolina the water temperature
gets rather low, but not quite so low as in New England. I
know that water temperature isn't the only factor that leads
to degeneration of the sponges in the fall because the North
Carolina sponge survives water temperatures much lower than
the water temperatures that occur when the sponges die back
in New England in the fall.

SIMPSON: There is one aspect of field observation, I think,
that hasn't been emphasized enough. When gemmules are forming and you can see them forming, or when gemmules are germinating, it is my guess that we should be going back much
further to find the trigger rather than looking for it around
that period when, for example, the swamps and the drainage
ditches are drying out.

DE NAGY: I've been doing some preliminary work with a non-
gemmulating sponge in New Hampshire. While it doesn't form
gemmules through the winter, there seems to be a marked reduction in flagellated chambers to the point of the sponge
being a solid mass of archeocytes by spring. This makes the
whole sponge essentially a resting stage without having produced a gemmule coat. Gemmules aren't the only solution to
overwintering.

HARRISON: The structure you describe is a reduction body.
This is another type of developmental system. It's similar
to the overwintering phase in terms of organization.

POIRRIER: I have produced reduction bodies in many species of freshwater sponge by simply putting them in habitat water in which gemmules had formed before. In other words, the sponge that had previously formed gemmules had apparently removed from the water silicon or trace elements that were needed for spicule production. When I put another sponge in this, generally I got the rapid production of reduction bodies, rather than typical gemmules.

SIMPSON: These are dishes in which vegetative tissue had been placed and then formed gemmules?

POIRRIER: Right.

Topic 7. What Do We Know About The Deposition Of Silicon In Sponges?

SIMPSON: I think one of the questions that is still un-answered is whether there is an infolding of the plasma membrane in the production of spicules. The electron microscopic work that Dr. Vaccaro and I have done has demonstrated that there can't be an infolding of the membrane for the whole length of the spicule or else in cross section a connection to the outside would be seen. The one possibility we couldn't eliminate is that there could be an inpocketing from the end of the cell so that only one neck opened out. This would mean that you would have to cut through the neck region to see that you're dealing with outside membrane. I think it's probably unlikely that that's the case because the axial filaments appear to be cytoplasmic structures. In the presence of germanium, it is even more strongly indicated that they are cytoplasmic structures. However, we haven't demonstrated that there isn't a final inpocketing of the cell membrane at the ends of the spicule.

HARRISON: I've just finished a study of siliceous shell formation in the testaceous ameba, Lesquereusia spiralis. In this ameba, it appears that silicon is concentrated in the Golgi apparatus in a manner somewhat similar to the formation of calcareous coccolith scales in the coccolithophorid algae. One thing we should look for in sponges is initial involvement of the Golgi apparatus, not only in the elaboration of actual spicule form, but also in the concentration of silicon itself. In micrographs of mature sponge silica-blasts, even with the Golgi apparatus showing, I have not seen the typical silicon electron density, but I have never been lucky enough to have an immature silicablast in the section. If Lesquereusia spiralis can be used as any reference, we can expect Golgi involvement in elaboration of silicon in sponges.

SIMPSON: There is, however, a major area of difference among the ameboid forms, the algae, and the diatoms. In sponges there is a whole membrane system which develops, the silicalemma. There is no indication at all of involvement of any silicon accumulation any place other than inside the silicalemma. This means that the silicon is pumped in, unpolymerized, and polymerization occurs after passage across the silicalemma.

The diatom group can synchronize the cells and look at the stages of silicon deposition. They have introduced the use of germanium in looking at silicon deposition, because germanium acts as a competitive inhibitor of silicon transport. The additional work that we've done with germanium in sponges is, I think, provocative because we've been able to indicate that germanium affects the growth in length of the organic axial filament, but apparently does not affect the surrounding silicalemma membrane. The spicules have a big bulb in the center. The germanium is inhibiting the growth of the organic filament in length, but is not inhibiting the surrounding silicalemma membrane, which continues to be produced, piles up, and balloons out. The end result is a spicule with one or more very large, bulbous structures. There can be either one central bulb, which is quite common, or a number of bulbous structures along the length. This indicates that the germanium is uncoupling the system. Normally, growth of the organic filament is coupled with the growth of the silicalemma, the surrounding membrane, to produce a geometric spicule. Germanium apparently uncouples them by inhibiting the growth of the filament but permitting the silicalemma to develop. Silicon is pumped into the silicalemma which flares out, producing the bulbous structure. Some of the spicules are nothing but a bulb with very short points.

POIRRIER: Dr. Harrison, do you know of any species of freshwater sponges which were described on the basis of spicular bulbs?

HARRISON: Yes. This is a classical malformation seen in freshwater sponges. If there is any one thing that is in-

dicative of an abnormal environment, it is spicules of
freshwater sponges with bulbous structures on them. This is
something one can always look for as an indicator of a dis-
turbed environment.

SIMPSON: It would be interesting to look at those environments in terms of germanium content. The diatom group working with Ben Volcani at Scripps have demonstrated, furthermore, that even though germanium is a competitive inhibitor
of silicon, it can actually be incorporated into the diatom
frustule. The frustule is hydrated amorphous silicon and
the germanium probably is also. The amazing thing is that
the germanium is deposited right into the frustule.

POIRRIER: Once I thought I had a new species of freshwater
sponge. It turned out to be Ephydatia fluviatilis. The
spicules were formed immediately after a heavy rain in a
brackish-water habitat,and there was a temporary thermocline.
I can show you these spicules, which have a series of bulbs
on them. The gemmoscleres, also, were malformed. An interesting thing is ecomorphic variation in Spongilla lacustris
in some of the very acid habitats in Louisiana, highly
colored habitats with an abundance of colored colloidal material in the water. If Spongilla lacustris is collected
from that habitat, the microscleres will be only 50 to 60 μ
in length and fairly thick. If the same sponge is placed in
the aquarium after the colored material precipitates out of
the water, the sponge will produce microscleres which are
120 μ in length and very, very slender. The same thing is
true of gemmoscleres. Although gemmoscleres are not produced in a very acid environment,they are produced in the
aquarium. This is again related to events occurring in the
water with the coloring colloidal material. The water is
acid in the natural habitat. When the colored material precipitates or is digested, the water becomes quite alkaline.
These observations, as related to the silicon cycle, would
make a very interesting study. Other things may be involved,
though, in terms of whether factors affect the silicon cycle
or the transport of this material across the cell membrane.

SIMPSON: Solubilities?

POIRRIER: Right. In other words, whether silicon is existing as a colloid or dissolved, or attached to particulate
organic material and transported in one of these states.

SIMPSON: There is little question that, when these bulbs

are produced, germanium uncoupled the systems. The membrane system continues producing membranes and still transports silicon, even though the filament remains short.

FELL: I have a question. In the natural habitat what incidence of these bulbous spicules do you find? Do you find that the majority or minority of spicules exhibits them?

POIRRIER: In species which live in brackish water, during times of osmotic stress with fluctuation of salinity,bulbous spicules will be produced.

FELL: Again, is that a small percentage of spicules?

POIRRIER: No. They could possibly be produced throughout the entire sponge. All the spicules which are produced during that period of stress will be bulbous. Those which existed prior to that period would not be.

SIMPSON: I think that there is something else that has to be emphasized. The very young microscleres of _Spongilla lacustris_ normally have a small central swelling. Occasionally immature megascleres do also. Some microscleres, even when they are a normal 120 to 130 μ in length, may show a small swelling. This probably indicates that in the microsclere, in terms of balance, the membrane is slightly ahead of the filament. There is no place for the silicon to go in a smooth geometric pattern, so it piles up to some extent.

POIRRIER: It is a beautiful explanation for what I have seen--even with the infolding.

SIMPSON: Normal swellings can occur, but it is necessary to distinguish between a swelling and a bulb. A bulb is a very large structure that is not normal.

POIRRIER: I tend to agree with that, too.

Topic 8. Problems of Sponge Cell Culture

CECIL: I have heard a lot about differentiation and aggregation of sponge cells and about culture of sponges. I assume that by the culture of sponges you mean the entire sponge. Is there any possible way in which any type of cell

43

in a sponge can be made to dedifferentiate? In other words,
can you take any type of cell and make it dedifferentiate
so that it will grow in continuous cell culture? I'm not
talking now about organotypic culture; I'm talking about a
single-cell culture. Can you do this through radiation?
I've read only one paper on the subject, which reported that
nothing happened with radiation. What about chemical carcinogens or any type of virus? Has any work been done on
this without resulting aggregation or differentiation?

LITTLE: Tracy Simpson may be more familiar with this than
I am, but I don't even think you can get sponge cell types
well separated.

SIMPSON: It's very difficult to separate them. The conclusion that one draws from reading the literature is that there
is very little that one could do to a sponge to get it to
culture.The system tends toward forming reduction bodies or
towards differentiation, but not towards flattened tissue
cultures.

CECIL: Developing culture of sponge cells would be a lot of
work but it seems that this would be very productive in terms
of chemotherapeutics or tumors.

SIMPSON: There is an unanswered question about whether
sponge cells will act as mammalian cells do and give homogenous cell types that are viable.

CECIL: You don't have to limit it to mammalian cells, because other invertebrate cells will do that, too, although
I'm not sure that they differentiate. However, they have
been grown in continuous culture.

LITTLE: Has anybody ever developed pure suspensions of one
sponge cell type? There is the problem.

SIMPSON: You can attempt to purify cell suspensions with
differential centrifugation, but you still get a fair amount
of contamination.

LITTLE: Dr. G. Evelyn Hutchinson commented on this same
problem six or seven years ago. He said, "Why don't you try
electrophoresis"?

CECIL: Is there anyone who believes that sponge cells are
not susceptible to chemical carcinogens? If not, why not?

HARRISON: Dr. G.P. Korotkova at Leningrad has subjected sponges to carcinogens. This is the only relevant paper that I know of.

SIMPSON: There is one reason to consider that it may not be possible to get sponges to proliferate in a pure culture condition. That is because most sponges genetically are capable of indefinite, but organized, growth. It may be that this balance is very, very tight in such a way that the sponge will go on proliferating, but only in a very specific manner. Many sponges are capable of growing with no set morphological limits, but they don't grow in an undefined pattern in the way cancerous tissue grows.

REISWIG: It seems that sponge cells are slightly different from other cells used in culture systems. The sponge cells have to be supplied by particulate food and this kind of food does not normally occur in a cell-culture system. They may also require the presence of an auxillary cell in the culture, perhaps an amebocyte carrier cell.

CECIL: You've got to have a contaminating system in other words?

REISWIG: I think that is the kind of thing we're going to have to look for.

MORALES: Excuse me, Dr. Reiswig. Are you implying that there is no transport through cell membranes in these organisms? It would seem that they could utilize some dissolved material.

REISWIG: Within the sponge I don't foresee any uptake of dissolved material except for informational flow, certainly not for trophic flow. This is apparently a necessary role of amebocytes in the sponges.

MORALES: Could it be that, up to this point, we haven't fed them exactly what they needed in terms of dissolved nutrient?

FELL: I think the potential problem that Dr. Reiswig has mentioned is a good one to keep in mind, but I think, also, that the amount of research on sponge cell separation and culture is very little in comparison to that devoted to systems which have been exploited in detail, like some of the mammalian systems and other systems. I don't think that results to date really indicate the probable inability to do this. I think that it simply means that we have to do a lot of work. More people have to go into this, in a sense, thankless type of work, to develop the proper types of culture media and cell separation techniques.

Topic 9. External Currents and Sponge Feeding

FROST: I'd like to raise a question about the work of Vogel, (eds. note; 1974, Biol. Bull., 147:443-456) in which he investigated the importance of external currents in relation to flow rates within sponges. In a laboratory study, using living and killed specimens of Halichondria bowerbanki and various mechanical models, he measured external currents and their effect on the rate of water flow out of natural and artificial oscula. He described an increase in outflow concomittant with an increase in external current and attributed this to the hydromechanics of the sponge feeding system, specifically the different sizes of the oscula and ostia and the pressures generated by different rates of water flow over these openings. I am interested in opinions about the importance of external current in generating what Vogel termed passive movement of water through the sponge. This is related to my studies of the feeding activity of Spongilla lacustris in which it is necessary to place a sponge in a chamber where the current flowing over the sponge could be significantly altered.

REISWIG: I think the hydrodynamics Vogel discusses are quite sound.

FROST: I certainly agree with the hydrodynamics. I was wondering about situations in natural populations of sponges. How important is this, say, in a lentic versus a lotic system?

STORR: There are many instances that I observed in which sediment was taken away from sponges in a small flow of water. This is partly sediment from the bottom, which is detrimental to the sponge itself, and the sponge has to backwash it. Also, of course, in a flow of water there is a much higher feeding rate. We are not talking about hydrodynamics per se, but about the amount of food that is coming to the sponge. We see enormous ecological variations in sponges from place to place. In one place the sponge will be quite

low with growth which is quite restricted. This sponge will
show, in essence, old age under poor ecological conditions.
In strong current, sponges will attain perhaps three to five
times greater size,and yet they will retain their normal
shape. An old sponge tends to have a doughnut shape.

FROST: A difficulty there, is to distinguish between direct
and indirect effects of external currents. By indirect, I
mean the physical transport of materials such as metabolites
and detritus from the area of the sponge, and current effects
making higher concentrations of food particles available. I
am interested in the direct effect of that current on the
filtering rate of the sponge. Vogel, through his artificial
system, addressed this. I was hoping to get at that critical
question.

REISWIG: Vogel used machined bottles of stainless steel,
drilled in dimensions approximating a millimeter or at least
one-half millimeter for the diameter of the confluent canals
through the system. This has no relationship at all to
sponges which have minimum dimensions of five microns entering flagellated chambers. The second system that Vogel
described involved insinuating a living sponge into a sand
vessel and measuring the change in flow rate across that
system. In no way did he ever validate or invalidate the
possibility that fluid flow was occurring through the sand,
out through the spongocoel, or the axial canal. The space
size between sand grains is four orders of magnitude larger
than the canals in the sponge. For these two reasons, I
think he has not demonstrated convincingly that fluid flow
is significantly augmented through the sponge by fluid flow
across the osculum. Certainly there is some necessary augmentation of fluid flow for any flow across the osculum as
there is some negative pressure developed, but I think this
negative pressure is, at most, 1% or 2% of the normal flow
through the sponge. I think that a change in fluid transported through the sponge would be negligible and probably
unmeasurable in the natural environment. This 1% or 2%,
is selectively advantageous for these sponges. This is apparently why most sponges in any moving environment are tubular and upright in shape or, if not upright, are oriented
perpendicular to the prevailing water current. I don't
think Vogel's conclusions are quantitatively significant,
but I think they are valid in concept.

Cell and Developmental Biology

ANALYSIS OF REPRODUCTION IN SPONGE POPULATIONS: AN OVERVIEW WITH SPECIFIC INFORMATION ON THE REPRODUCTION OF HALICLONA LOOSANOFFI[1]

Paul E. Fell

Department of Zoology
Connecticut College
New London, Connecticut 06320

ABSTRACT: In populations of sponges, which have internal embryonic and larval development (most sponges), reproductive activity may be analyzed by the following methods: microscopic examination of tissue samples, observation of larval release, sampling larvae in the plankton, and study of larval settlement on natural and experimental surfaces. These methods have been used to study reproduction in Haliclona loosanoffi and to some extent in a species of Halichondria. In the Mystic Estuary, Connecticut, many specimens of Haliclona loosanoffi contain large numbers of gametes and/or embryos from late May, when the sponges develop from gemmules, through the middle of July. Specimens with some reproductive elements may also be found in late July and August. This sponge appears to be gonochoric with a sex ratio of approximately 1:1. Larval release and settlement take place from about the second week in June through about the first week in August, and the peak of larval settlement occurs in July. Larval behavior seems to play an important role in the pattern of larval settlement. Although sampling larvae in the plankton has not proved to be successful with Haliclona loosanoffi, preliminary studies suggest that this may be a useful method for studying reproduction in Halichondria. In Haliclona loosanoffi gemmules begin to develop in late June and by the second week in August all of the specimens possess these structures. When refrigerator-stored gemmules are implanted back into the natural habitat of the sponge, gametes, embryos, larvae, and gemmules may develop within a period of one month (at ca 25°C) or within about 3 weeks after the development of small sponges.

As pointed out by Diaz ('73), Simpson and Gilbert ('73),

---
[1]The original work reported here was supported by a Cottrell College Science Grant from the Research Corporation.

51

Fell ('74a) and others, there have been few detailed studies on the reproduction of sponges. Since reproduction is of such fundamental importance, it is desirable that information on this process and its regulation be accumulated. In the past few years interest in this area of sponge biology has grown, and it is not unreasonable to expect that within the next decade substantial advances in our knowledge will be made. Not only is information on reproduction of considerable inherent interest, but some of it can also be valuable in attempting to resolve taxonomic questions.

Sponge life histories

Some sponges are typically annuals, the active form of the sponges existing for only a portion of the year. Obviously sexual reproduction of such sponges is restricted to a definite period, and this period is frequently shorter than the active period of the sponges. Most fresh-water and some marine species occur only as resistant gemmules during certain periods (see Fig. 1). However, the dormant period may be at different times of the year in different parts of the geographical range of a sponge and even in different habitats within a restricted region (Poirrier, '69; Wells, Wells and Gray, '64; Fell, '74b; Simpson and Fell, '74). In addition, some sponges may form gemmules in certain habitats but not in others (Poirrier, '69). A few marine sponges, which do not produce gemmules, regress during the winter to a simplified form of reduced size lacking flagellated chambers (Hartman, '58; Simpson, '68; also see Stone, '70). The dormant form of these sponges is analogous to gemmules in many respects. Even when there is no true dormancy the rate of growth and presumably the rates of other processes may vary seasonally, especially in temperate regions (see Wells, Wells and Gray, '64).

Most marine sponges remain more or less active throughout the year, and some of them may live for many years (Reiswig, '73; Dayton, Robilliard, Paine and Dayton, '74). In such species reproduction may either be restricted to a definite period(s) of time at any particular location or occur continuously. In the latter case the level of reproductive activity may vary throughout the year (Storr, '64).

Methods of studying reproduction

A recent review of the most widely used methods for studying reproduction in animals has been provided by Giese and Pearse ('74). A discussion of how some of these methods

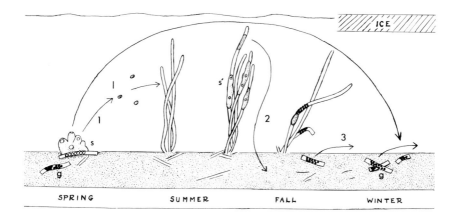

Fig. 1. Life history of <u>Haliclona loosanoffi</u> in the Mystic
Estuary, Connecticut. During the colder months of the year
this sponge exists only as gemmules (g); these germinate in
late spring forming new sponges (s) which soon begin to reproduce sexually and produce planktonic parenchymula larvae
(1); the larvae settle on eel grass and other substrates
where they undergo metamorphosis into small sponges (s′).
During the summer both the gemmule-derived and larva-derived
sponges grow and produce gemmules; in the early autumn the
sponges die leaving the gemmules attached to the substrate.
Dispersal of the sponge is effected by means of larvae (1),
sponges attached to fragments of eel grass and algae (2),
and gemmules (3).

have been applied in the study of reproduction of sponges is
presented here. In populations of sponges, which have internal embryonic and larval development (most sponges), reproductive activity may be analyzed by the following methods: microscopic examination of tissue samples, observation
of larval release, sampling larvae in the plankton, and
study of larval settlement on natural and experimental surfaces. Observation of gamete release by sperm producing
specimens and by oviparous egg producing specimens may also
yield valuable information concerning reproduction. In few
cases have any of these methods been fully exploited. It is
desirable that a combination of methods be used whenever it
is practicable.

Examination of tissue samples

Microscopic examination of tissue samples is the method of analysis which is capable of yielding the greatest amount of information about the reproductive process. Not only is it possible to determine the reproductive period and to estimate reproductive activity by this means, but information concerning the form of sexuality, gametogenesis, and in most cases embryonic and larval development may also be obtained (see Siribelli, '62; Simpson, '68; Liaci and Scisciolo, '67 and '70; Simpson and Gilbert, '73 and '74; Diaz, '73; Reiswig, '73; and Chen, in this volume).

In many cases, when the embryos and larvae are relatively large, the reproductive period can be determined by the examination of cut up specimens with a dissecting microscope (or even with the naked eye). This method is useful when large numbers of specimens are to be studied (Storr, '64; Stone, '70); but unless some of the specimens are subsequently examined histologically, much valuable information is lost. Even when a detailed histological study is undertaken, it is frequently desirable that a thorough examination of "dissected" specimens under a dissecting microscope be made. For example, it sometimes happens that a few embryos are restricted to a small portion of a specimen, and these might not be included in a random sample taken for histological study.

The study of histological sections has been used to obtain valuable qualitative information concerning reproductive activity (Simpson, '68; Fell, '70; Simpson and Gilbert, '73 and '74; Reiswig, '73). However, quantitative studies should be made whenever possible. In a number of cases specimens containing reproductive elements have been reported to occur over an extended period of time, even throughout the year, while a high level of reproductive activity may be restricted to a much shorter period (Storr, '64; Fell, '70 and '74a).

Data from studies of tissue samples may be expressed quantitatively as the percentage of specimens containing reproductive elements during a particular period (Siribelli, '62; Storr, '64; Liaci and Scisciolo, '67 and '70; Stone, '70) and/or as the percentage of specimens possessing large numbers of reproductive elements (see Fig. 2). It is also possible to enumerate reproductive elements in tissue sections of known area and thickness in order to estimate levels of reproductive activity within individual specimens and to obtain mean values and ranges for the population at any particular time. Such enumeration is done best under oil immersion at a magnification of about 1000 X.

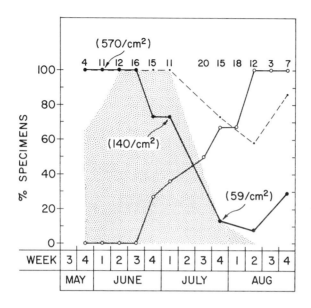

Fig. 2. Reproductive activity of Haliclona loosanoffi in
the Mystic Estuary, Connecticut. Sexual reproduction: the
large solid circles joined by a heavy line represent the
percentages of specimens with large numbers of reproductive
elements; the dots joined by a broken line indicate the percentages of specimens containing at least some reproductive
elements; the numbers in parentheses are mean numbers of
reproductive elements per cm^2 histological section (10 µm
thick) in female specimens; and the stippled area shows the
percentages of female specimens with large oocytes, embryos
and/or larvae. Asexual reproduction: the large open circles
joined by a heavy line represent the percentages of specimens containing gemmules. The number of specimens sampled
during each period is indicated at the top of the graph.
Note the apparent antagonism between gemmule formation and
sexual reproduction.

There is relatively little information available on the
sexuality of sponges. Contemporaneous hermaphrodism has
been shown to occur in a number of species (Tuzet, '32;
Meewis, '38; Lévi, '53 and '56; Sarà, '61; Simpson, '68;
Reiswig, '73), while a separation of the sexes has been
found to exist in others. Gonochorism appears to occur in

many of the sponges in which the sexes are separate (Egami and Ishii, '56; Tuzet and Pavans de Ceccatty, '58; Sarà, '61; Sirribelli, '62; Liaci and Scisocioli, '67 and '70; Simpson and Gilbert, '73); however, a distinction between gonochorism and successive hermaphrodism is not always easy to make (Diaz, '73). Although in a number of instances the latter form of sexuality has been suggested (see Fell, '74a), in only a few cases has it been clearly demonstrated (Sarà, '61; Diaz, '73). The best method for distinguishing between gonochorism and successive hermaphrodism is analysis of tissue samples taken from the same specimens over an extended period of time (Diaz, '73; Gilbert, '74). It is desirable in such studies both to examine successive samples taken from the same specimens and to study entire specimens collected at different times during the reproductive period. To facilitate repeated sampling of specimens from mirky waters and/or habitats which provide other difficulties in relocating particular sponges, the specimens may be suspended in cages from floats or other structures (Diaz, '73).

Too few specimens of any sponge species have been adequately studied to determine sex ratios for species in which the sexes are separate. However, in a few cases rough estimates may be made (see Table 1). In order to estimate the sex ratio, one must select a portion of the reproductive period when nearly all of the sponges are reproductive. For example, in some cases sperm do not develop until after the oocytes have nearly completed their growth (Siribelli, '62; Liaci and Sciscioli, '67), and in all cases there is no way of being sure of the sex of non-reproductive specimens. More studies in this area need to be made using larger numbers of specimens.

The literature on gametogenesis in sponges has been recently reviewed by Fell ('74a) and that on embryonic development by Borojević ('70) and Fell ('74a). Additional information on these subjects may be found in many of the references included in this paper. In many sponges a developmental sequence, including gametogenesis, early and late cleavage, and larval development, can be readily discerned. However, in a number of species the developmental sequence is less obvious. Frequently difficulty in establishing a typical sequence is due, at least in large part, to a great abundance of reserve substances which obscure cytological details. In a few instances this difficulty has led to the erroneous conclusion that development of larvae is by an asexual process (see Gureeva, '72). Although it was recently reported that a number of intertidal sponges produce larvae asexually (Bergquist, Sinclair and Hogg, '70), it now

TABLE I

THE RELATIVE ABUNDANCE OF MALE AND FEMALE SPECIMENS
IN POPULATIONS OF SOME SPONGES

Species	No. reproductive specimens examined[1]	Approx. ratio male:female	Author
Stelletta grubii	51	1:3	Liaci & Sciscioli,1970
Haliclona ecbasis	57	$1:2^2$	Fell, 1970
Erylus discophorus	104	1:1.5	Liaci & Sciscioli,1970
Haliclona loosanoffi	70	1:1	Fell, in prep.
Hymeniacidon sanguinea	65	4.5:1	Sarà, 1961

[1]During period when at least most of the specimens were reproductive.
[2]In addition to the males and females, there were 5 hermaphrodites.

appears that this interpretation is incorrect (Bergquist, personal communication). Serial section analysis of developmental stages can be very useful in documenting a course of development which might otherwise be difficult to establish (see Fell, '69 and Fig. 3). Although this type of analysis is very time consuming, it should be used more widely in studies of sponge development.

Microscopic examination of tissue samples may also be used to study asexual reproduction by means of gemmules (Simpson and Gilbert, '73 and '74; Fell, '74b) and buds (Devos, '65).

Observation of the release of larvae (and/or gametes)

The period(s) during which larvae are released may be studied either by bringing specimens collected at frequent intervals into the laboratory for short periods of time (6 to 24 hrs) to see if they release larvae (Fell, in preparation) or by maintaining specimens in the laboratory for long periods of time and observing them at frequent intervals (Fry, '71). It is desirable in all cases that intact specimens be used in order to assure, as far as possible, against the release of incompletely developed larvae. This means that such studies are done best with sponges which are either simply

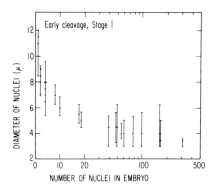

Fig. 3. Demonstration of cleavage in the embryos of
Haliclona ecbasis by serial section analysis. Although this
process is not obvious upon casual observation, enumeration
and measurement of nuclei in serial sections of the embryos
show convincingly that it occurs. The means and ranges of
nuclear diameters in embryos containing various numbers of
nuclei are shown. Beyond the 18-nucleus stage only 15 to 20
nuclei were measured in each embryo.

resting on the bottom or attached to substrates which can be
taken back to the laboratory. Study of the larvae following
their release can yield valuable information concerning
larval behavior and longevity, substrate preferences, and
metamorphosis (see Warburton, '66; Bergquist and Sinclair,
'68; Simpson, '68; Fry, '71; Jones, '71). A detailed review of factors influencing settlement of marine invertebrate larvae is provided by Crisp ('74).

In regions where there is good visibility under water, it
may be possible to observe in the field the release of large
numbers of larvae by large specimens of certain species.
This has not yet been reported; but as more detailed studies
of sponge populations are executed by diving, such observations may be made.

Theoretically one should be able to study the release of
sperm by specimens of all sexually reproducing species and
observe the release of oocytes or zygotes by specimens of
oviparous forms. As in the case of the release of larvae,
these studies may be conducted in the laboratory (Lévi, '51
and '56; Watanabe, '57; Warburton, '58; Borojević, '67
and/or in the field (Reiswig, '70 and in this volume).

Sampling larvae in the plankton

Sampling larvae in the plankton has not been previously used to analyze reproduction in sponge populations, although there is no reason why this method should not be used in certain favorable situations. The density of sponge larvae in the plankton is probably rarely (if ever) high, making it difficult to perform meaningful quantitative studies. However, qualitative studies may be used to document the presence of sponge larvae in the plankton during a particular period(s). In an analysis of the reproduction of a species of Halichondria, plankton sampling ahs provided information which correlates well with data obtained by other methods (Fell and Jacob, in preparation).

Study of larval settlement

The study of larval settlement on experimental surfaces may be used to obtain quantitative data on reproductive activity (see Fig. 4). Since most sponge larvae spend a very short

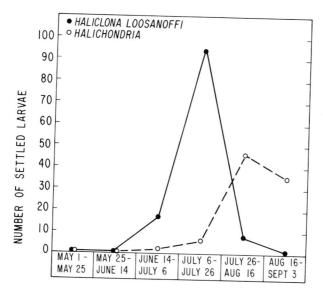

Fig. 4. Settlement of the larvae of Haliclona loosanoffi and a species of Halichondria on Mercenaria shells suspended in the lower Mystic Estuary, Connecticut. Ten fresh shells were set out during each period.

time (1-2 days) in the plankton, this method gives indirect

information concerning the time of larval development and release.

Various materials on which sponge larvae settle may be
placed in the water below and/or within the levels of tidal
exposure (in the case of marine species). Among the materials which have been used successfully are mollusc shells,
unglazed tiles, rocks and plexiglass. These may be suspended
in trays or cultch bags, strung on cord, or attached to submerged timbers. In some cases a number of different materials may be employed in order to gain information on possible
substrate preferences (see McDougall, '43; Hartman, '58;
Wells, Wells and Gray, '64; Simpson and Gilbert, '74).

Generally three methods of exposure of settling surfaces
have been used: 1) exposing the same surfaces for the entire
study period and examining them at regular intervals to determine how many larvae of each species settled during each
interval (McDougall, '43; Hartman, '58); 2) placing materials in the water at regular intervals to remain there until
the end of the study and examining them frequently as in
method 1 (Hartman, '58); 3) setting out fresh surfaces during
each observation interval (McDougall, '43; Wells, Wells and
Gray, '64). In the last two methods there is the advantage
of being able to look at larval settlement on similar surfaces during each interval. When materials are submerged in
the water for even short periods of time their surfaces are
altered by the development of films (Crisp, '74); and as more
organisms progressively colonize these surfaces, there is
less space available for the settlement of new larvae. In
all cases the surfaces should be examined for the presence of
small sponges under a dissecting microscope (Wells, Wells and
Gray, '64). When the young postlarvae of two or more species
are difficult to distinguish, it is desirable that the identification of each small sponge be checked by preparing a wet
mount of it and examining the spicules at a magnification of
at least 400 X.

There are two potential problems associated with this method of analysis. First, the larvae may experience different
levels of competition for settling space during different
intervals, so that one may not get a true picture of larval
production with time, although one may get an accurate estimate of relative settling success. Ascidians, such as
Molgula and Botryllus, as well as other organisms which settle in large numbers and grow rapidly may dominate settling
surfaces for certain periods (Wells, Wells and Gray, '64).
Secondly, it is not always clear where the larvae that settle come from. Larvae may be carried into and out of the
study area by currents (Carriker, '59; Stone, '70). However,

since the larvae are free-swimming for only a short time, in many cases they probably are not carried very far from where they are released before they settle.

The study of larval settlement on natural surfaces, such as eel grass, algae and bare rock surfaces, may also yield important, but frequently somewhat more limited, information on reproduction. Plants which undergo regrowth each spring are well suited to some such studies since they do not support sponge specimens persisting from the previous year. Rock surfaces may be cleared of existing organisms with a chisel and stiff wire brush (Stone, '70).

Growth and/or reproduction of sponge postlarvae have been studied in only a few cases (see Hartman, '58; Reiswig, '73; Simpson and Gilbert, '74; and Fig. 5), and it is desirable that more studies of this sort be made. Not only is it of interest to know whether there are 2 or more generations undergoing sexual reproduction within a single season or year, but such studies may also lend some insights into what factors are important in regulating reproductive activity.

Regulation of reproduction

Basically two approaches have been used in studying the regulation of reproductive activity in sponge populations: 1) to study reproduction of particular sponge species at several locations within their geographical ranges in an attempt to correlate reproductive activity with environmental factors and 2) to study sponges in the laboratory or manipulate them within their natural environment in an effort to determine the importance of various environmental factors in regulating the reproductive process. Both approaches have been used to a very limited extent.

The first type of study may be carried out by a single person or group working at several locations or by different individuals working in different regions. In the latter situation efforts should be made to make the separate studies as comparable as possible. In Gulf of Mexico populations of Hippiospongia lachne the peak of reproductive activity is reached earliest at British Honduras, later at the Bahamas and latest at Cedar Keys, Florida following the rise in water temperature. The optimal temperature range for reproduction appears to be between 23 and 29°C (Storr, '64). Similarly, the reproduction of Haliclona loosanoffi has been studied at Milford, Connecticut (Hartman, '58), Mystic, Connecticut (Fell, '74b and in preparation) and Hatteras Harbor, North Carolina (Wells, Wells and Gray, '64); and it was found that although larval settlement begins at different times at these

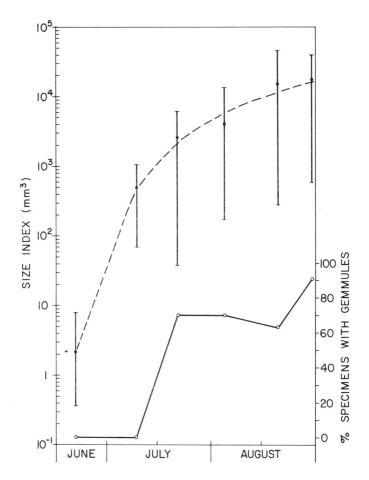

Fig. 5. Growth of and gemmule production by postlarvae of
Haliclona loosanoffi in the Mystic Estuary, Connecticut.
The size index is a product of length X width X height.
Although some of these specimens contained a few small
oocytes or spermatic cysts, none contained embryos or lar-
vae.

locations, it always begins about one month after the water
temperature reaches approximately 20°C. Other species, the
reproduction of which has been studied in detail in two or
more regions, include Axinella damicornis (Siribelli, '62;
Lévi, '51) and Microciona prolifera (Hartman, '58; Wells,
Wells and Gray, '64, Simpson, '68). Not only are more

studies on the reproduction of various species at different locations needed, but it is also important that attempts be made to standardize the methods of study and ways of reporting data so that meaningful comparisons can be made.

Studies of reproduction under laboratory conditions, in which some environmental variables may be changed indepen-

Prior to 1984

entially capable of yielding im-
e regulation of reproduction but
lems. In only a few cases have
initiate or sustain reproduction in
only relatively few species have
)oratory for extended periods of
Under some conditions explants of
a mixture of algae, continue to
y for about 3-4 weeks (Simpson,
duced and released at 20 and 25°C,
even though growth is normal at
arvae form at 10°C. Specimens of
e been maintained in the labora-
d bacteria for from 2 to 3 months,
ae occurred during the same period
to be reproductive in the field
the sponges were in the labora-

tory for more than a month before reproduction (as judged by the release of larvae) was initiated. More such attempts to study reproduction in the laboratory need to be made.

The dormant forms of certain sponges provide a useful means of manipulating them within their natural habitat. For example, gemmules may be stored at low temperature for relatively long periods of time and then allowed to germinate in their natural environment at times of the year which differ from the normal period fo gemmule germination (Gilbert, '74). When refrigerator-stored gemmules of _Spongilla lacustris_ are implanted back into the pond of origin at different times during the spring and summer, they germinate normally forming new sponges. Although the normal reproductive period of this species in the region of study (New Hampshire) is the late spring, newly formed female sponges begin to produce small oocytes within one week even when implantation is delayed until late summer. The oocytes develop; and if male specimens are present, embryos and larvae are produced. These facts suggest that environmental conditions are favorable for reproduction during a period of time which is substantially longer than the normal reproductive period and that reproductive activity is also regulated by endogenous factors (Gilbert, '74). Similar studies should be carried out with other gemmuliferous sponges. Gemmules may also be

implanted into different habitats in an attempt to evaluate
the effects of various environmental factors on reproduction.
Finally, gemmule implants may be used to obtain more precise
information on the time required for various processes within the reproductive sequence (i.e. oogenesis, embryonic development, etc.).

The overwintering form of Halichondria sp. and presumably
that of other sponges (see Hartman, '58; Simpson, '68) can
be stored at low temperature and handled in the same way as
gemmules (Fell, in preparation). Therefore the type of experiments done with Spongilla lacustris in the field
(Gilbert, '74) should also be possible with some of these
sponges which do not gemmulate.

ACKNOWLEDGMENTS

The author wishes to thank Dr. Frances Roach, Ruth Fell,
Sibyl Hausman, and Jeanie Kitchen for expert assistance with
the original work presented here and Mrs. John Faigle and
Mrs. Robert Jacques for help with the Italian translation.

LITERATURE CITED

Bergquist, P.R. and M.E. Sinclair 1968 The morphology and
   behaviour of larvae of some intertidal sponges. New Zeal.
   J. Mar. and Freshwat. Res., 2: 426-437.
Bergquist, P.R., M.E. Sinclair, and J.J. Hogg 1970 Adapta-
   tion to intertidal existence: reproductive cycles and lar-
   val behaviour in Demospongiae. In: Biology of the
   Porifera, Sym. Zool. Soc. Lond. No. 25 (W.G. Fry, ed.),
   Academic Press, New York, pp. 247-271.
Borojević, R. 1967 La ponte et le développement de
   Polymastia robusta. Cah. Biol. Mar., 8: 1-6.
_____1970 Différenciation cellulaire dans l'embryogenèse et
   la morphogenèse chez les spongiaires. In: Biology of the
   Porifera, Sym. Zool. Soc. Lond. No. 25 (W.G. Fry, ed.),
   Academic Press, New York, pp. 467-490.
Carriker, M.R. 1959 The role of physical and biochemical
   factors in the culture of Crassostrea and Mercenaria in a
   salt-water pond. Ecol. Monographs, 29: 219-226.
Crisp, D.J. 1974 Factors influencing settlement of marine
   invertebrate larvae. In: Chemoreception in Marine Organ-
   isms. (P.T. Grant and A.M. Mackie, eds.) Academic Press,
   New York, pp. 177-265.
Dayton, P.K., G.A. Robilliard, R.T. Paine, and L.B. Dayton
   1974 Biological accommodation in the benthic community at
   McMurdo Sound, Antarctica. Ecol. Monographs, 44: 105-128.

Devos, C. 1965 Le bourgeonnement externe de l'éponge
Mycale contarenii (Martens) (Demosponges). Bull. Mus.
Nation. Hist. Natur., 37: 548-555.
Diaz, J.-P. 1973 Cycle, sexuel de deux demosponges de
l'étang de Thau: Suberites massa Nardo et Hymeniacidon
caruncula. Bowerbank. Bull. Soc. Zool. France, 98: 145-
156.
Egami, N. and S. Ishii 1956 Differentiation of sex cells
in united heterosexual halves of the sponge, Tethya
serica. Annot. Zool. Japan, 29: 199-201.
Fell, P.E. 1969 The involvement of nurse cells in oogenesis and embryonic development in the marine sponge,
Haliclona ecbasis. J. Morphol., 127: 133-150.
_____ 1970 The natural history of Haliclona ecbasis de
Laubenfels, a siliceous sponge of California. Pacific
Sci., 24: 381-386.
_____ 1974a Porifera. In: Reproduction of Marine Invertebrates. (A.C. Giese and J.S. Pearse, eds.), Vol. I, Aca-
demic Press, New York, pp. 51-132.
_____ 1974b Diapause in the gemmules of the marine sponge,
Haliclona loosanoffi, with a note on the gemmules of
Haliclona oculata. Biol. Bull., 147: 333-351.
Fry, W.G. 1971 The biology of larvae of Ophlitaspongia
seriata from two North Wales populations. In: Fourth
European Mar. Biol. Sym. (D.J. Crisp, ed.) Cambridge U.
Press, pp. 155-178.
Giese, A.C. and J.S. Pearse 1974 Introduction: general
principles. In: Reproduction of Marine Invertebrates.
(A.C. Giese and J.S. Pearse, eds.), Vol. I, Academic
Press, New York, pp. 1-49.
Gilbert, J.J. 1974 Field experiments on sexuality in the
freshwater sponge Spongilla lacustris: the control of
oocyte production and the fate of unfertilized oocytes.
J. Exp. Zool., 188: 165-178.
Gureeva, M.A. 1972 "Sorites" and oogenesis in the Baical
endemic sponges. Citologije, 14: 32-45.
Hartman, W.D. 1958 Natural history of the marine sponges
of southern New England. Bull. Peabody Mus., Yale, 12:
1-155.
Jones, W.C. 1971 Spicule formation and corrosion in re-
cently metamorphosed Sycon ciliatum (O. Fabricius). In:
Fourth European Mar. Biol. Sym. (D.J. Crisp, ed.)
Cambridge U. Press, pp. 301-320.
Lévi, C. 1951 Remarques sur la faune des spongiaires de
Roscoff. Arch. Zool. Exp. Gén., 87: 10-21.

_____1953 Déscription de <u>Plakortis</u> <u>nigra</u> nov. sp. et re-
marques sur les Plakinidae (Démosponges). Bull. Mus.
Hist. Nat., Paris, <u>25</u>: 320-328.
_____1956 Étude des <u>Halisarca</u> de Roscoff: embryologie et
systématique des Démosponges. Arch. Zool. Exp. Gén., <u>93</u>:
1-181.
Liaci, L.S. and M. Sciscioli 1967 Osservazioni sulla
maturazione sessuale di un Tetractinellide: <u>Stelletta</u>
<u>grubii</u> O.S. (Porifera). Arch. Zool. Italy, <u>52</u>: 169-177.
_____1970 Il ciclo sessuale di <u>Erylus</u> <u>discophorus</u> (Schmidt)
(Porifera Tetractinellida). Riv. Biol., <u>63</u>: 255-263.
McDougall, K.D. 1943 Sessile marine invertebrates of
Beaufort, N.C. Ecol. Monographs, <u>13</u>: 321-374.
Meewis, H. 1938 Contribution á l'étude de l'embryogénèse
des Myxospongidae: <u>Halisarca</u> <u>lobularis</u> (Schmidt). Arch.
Biol., <u>50</u>: 3-66.
Poirrier, M.A. 1969 Louisiana fresh-water sponges: tax-
onomy, ecology, and distribution. Ph.D. Thesis, Louisiana
State University, Univ. Microfilms Inc., Ann Arbor, Mich.,
No. 70-9083, 173 p.
Rasmont, R. 1961 Une technique de culture des éponges
d'eau douce en milieu controlé. Ann. Soc. Roy. Zool.
Belg., <u>91</u>: 147-156.
Reiswig, H.M. 1970 Porifera: sudden sperm release by
tropical Demospongiae. Science, <u>170</u>: 538-539.
_____1973 Population dynamics of three Jamaican Demo-
spongiae. Bull. Mar. Sci., <u>23</u>: 191-226.
Sarà, M. 1961 Ricerche sul gonocorismo ed ermafroditismo
nei Porifera. Boll. Zool., <u>28</u>: 47-60.
Simpson, T.L. 1968 The biology of the marine sponge
<u>Microciona</u> <u>prolifera</u> (Ellis and Solander). II. Tempera-
ture-related, annual changes in functional and reproduc-
tive elements with a description of larval metamorphosis.
J. Exp. Mar. Biol. Ecol., <u>2</u>: 252-277.
Simpson, T.L. and P.E. Fell 1974 Dormancy among the
Porifera: gemmule formation and germination in fresh-water
and marine sponges. In: Perspectives on the Biology of
Dormancy. (J.H. Bushnell, ed.) Trans. Amer. Microscop.
Soc., <u>93</u>: 544-577.
Simpson, T.L. and J.J. Gilbert 1973 Gemmulation, gemmule
hatching and sexual reproduction in fresh-water sponges.
I. The life cycle of <u>Spongilla</u> <u>lacustris</u> and <u>Tubella</u>
<u>pennsylvanica</u>. Trans. Amer. Microscop. Soc., 422-433.
_____1974 Gemmulation, gemmule hatching, and sexual repro-
duction in fresh-water sponges. II. Life cycle events in
young larva-produced sponges of <u>Spongilla</u> <u>lacustris</u> and an

unidentified species. Trans. Amer. Microscop. Soc., 93: 39–45.

Siribelli, L. 1962 Differenze nel ciclo sessuale di Axinella damicornis (Esper) ed Axinella verrucosa O. Sch. (Demospongiae). Boll. Zool., 29: 319–322.

Stone, A.R. 1970 Growth and reproduction of Hymeniacidon perleve (Montagu) (Porifera) in Langstone Harbor, Hampshire. J. Zool., G.B., 161: 443–459.

Storr, J.F. 1964 Ecology of the Gulf of Mexico commercial sponges and its relation to the fishery. Spec. Sci. Rep. U.S. Fish Wildl. Serv., Fish No. 466, 73 p.

Tuzet, O. 1932 Recherches sur l'histologie des éponges Reniera elegans (Bow) et Reniera simulans (Johnston). Arch. Zool. Exp. Gén., 74: 169–192.

Tuzet, O. and M. Pavans de Ceccatty 1958 La spermatogenèse, l'ovogenèse, la fécondation et les premiers stades du développement d'Hippospongia communis L.M.K. (=H. equine O.S.). Bull. Biol., 92: 331–348.

Warburton, F.E. 1958 Reproduction of fused larvae in the boring sponge Cliona celata. Nature, 181: 493–494.

_____ 1966 The behavior of sponge larvae. Ecology, 47: 672–674.

Watanabe, Y. 1957 Development of Tethya serica Lebwohl, a tetraxonian sponge. I. Observations on external changes. Nat. Sci. Rep. Ochanomizu Univ., 8: 97–104.

Wells, H.W., M.J. Wells, and I.E. Gray 1964 Ecology of sponges in Hatteras Harbor, North Carolina. Ecology, 45: 752–767.

CYTOCHEMICAL STUDIES OF CONNECTIVE TISSUES IN SPONGES

Ronald R. Cowden

School of Medicine
East Tennessee State University
Johnson City, Tennessee 37601

and

Frederick W. Harrison

Department of Anatomy
Albany Medical College
Albany, New York 12208

ABSTRACT: A variety of cytochemical methods were used to characterize the connective tissues of a series of sponges. Basic chemical differences may exist between connective tissue components, even within the same species of sponge. All sponge connective tissues contain high levels of protein disulfide groups. It is suggested that sponge collagen is chemically similar to the mammalian collagen precursor, procollagen.

In recent years a number of physicochemical and ultrastructural studies (Pavans de Ceccatty and Thiney, '63; Borojevic and Levi, '67; Garrone, '69; Garrone and Pavans de Ceccatty, '71; Vacelet, '71; Huc and de Vos, '72; Garrone et al., '73; de Vos, '72; de Vos and Rozenfeld, '74; Ledger, '74) have confirmed and extended the earlier studies of Marks et al. ('49) and Gross et al. ('56) which demonstrated the collagenous nature of sponge connective tissue fibers. It is now considered that there are two types of fibrous components in sponge connective tissue. The first, "Spongin A" of Gross et al. ('56) is composed of intercellular collagen fibrils generally less than 80 Angstroms in diameter, often with clear periodicity and typically coursing throughout the mesohyl. Junqua et al. ('74) define these fibrils as being only visible by electron microscopy. The second fibrous component, "Spongin B" of Gross et al. ('56), is composed of large anastomosing fibers, 10-15 microns wide, which although collagenous in nature, are structurally unique to sponges. Because of their structural specificity and

selective occurence, Garrone ('69), Vacelet ('71), and Junqua ('74) have suggested that they be identified by the term
"Spongin B" or, simply, spongin. "Spongin A", these authors
suggest, should be simply designated collagen or intercellular collagen because of its close relationship to collagen
fibers of higher metazoa.

Histological and cytochemical studies on connective tissue
in sponges have been carried out on a relatively limited number of organisms. In most cases (Simpson, '68) this work has
been incidental to other studies. Cowden ('70) employed a
variety of histological and absorption cytochemical methods
to document the complexity and variation of organization of
connective tissue or scleroprotein exoplasms in a series of
Gulf and Caribbean demosponges.

The present study using sponge specimens selected for their
differing complexity of organization, complements and extends
earlier studies of sponge connective tissues and exoplasms
through the extensive application of fluorescent cytochemical
methodology, particularly those techniques demonstrating
specific macromolecular components or reactive end-groups.

MATERIALS AND METHODS

The "HI" series of specimens was collected at Heron Island,
Great Barrier Reef, Queensland, Australia. These specimens
were field-fixed in either Bouins solution or ethanol:acetic
acid (3:1). The spongillid, _Ephydatia_ _fluviatilis_, was fixed
in ethanol:acetic acid (3:1). Specimens were routinely paraffin embedded and were sectioned at 5 mμ.

Masson's connective tissue sequence using aniline blue as
the collagen staining component was used on all specimens.
Sections were routinely stained with the periodic acid-Schiff
(PAS) method following prior diastase extraction to demonstrate polyvicinal glycols, principally neutral mucins. Similarly fixed tissues were subjected to Alcian blue 8GX staining at pH 1 and pH 2.5 for demonstration of sulfomucins and
less acidic, principally carboxylated, mucins respectively
(Mowry, '63). Basic groups, ionizable amino groups, were
stained with biebrich scarlet, pH 2.8, according to principles developed by Deitch ('55).

A variety of fluorescent cytochemical techniques were used
on ethanol-acetic acid (3:1)--fixed tissues for demonstration of macromolecular components or specific end-groups.
The dansylchloride procedure (Rosselet and Ruch, '68) was
used for localization of lysine. The modified Morel-Sisley
method of Ritter and Berman ('63) was used in the fluorescence mode according to Harrison et al. ('75) for demonstra-

70

tion of tyrosine. Brilliant sulfoflavine (BSF), pH 2.8, was used to fluorochrome ionizable amino groups according to principles developed by Leeman and Ruch ('72). The salicyl-hydrazide-zinc method (Stoward and Burns, '67) was used for demonstration of C-terminal carboxyl groups. Side-chain carboxyl groups were fluorochromed using the hydroxy-naphthoic acid hydrazide (HNAH) method of Curtis and Cowden ('71). The fluorescein mercuric acetate (FMA) method (Cowden and Curtis ('70) was used for protein sulfhydryl (SH) groups. The same method after thioglycollate reduction was employed to demonstrate the sum of SH plus disulfide (SS). Rigler's ('66) acridine orange method for demonstration of DNA and RNA as well as control preparations pretreated with RNase (Swift, '66) were also used on ethanol-acetic acid (3:1)--fixed tissues.

Preparations were mounted in most instances in Harleco fluorescence-free mounting media. The HNAH preparations were mounted in Gurr's "hydramount" and the acridine orange preparations were mounted in buffer and sealed. Photographs were made using fluorescence-free immersion oil.

All fluorescent preparations were observed and photographed using a Zeiss Photomicroscope II with an incident FL II fluorescence condenser. The high intensity lamp housing contains a built-in BG-38 heat absorbing filter. A 150 watt Xenon arc source was employed for excitation in combination with a BG-12 primary filter and a 530 nm secondary filter. Kodak Plus-X film was used for photography. "Diafine" development allowed the film to be used at an effective A.S.A. of 640.

OBSERVATIONS

Tissues of the Heron Island specimens _Geodia globostelli-fera_ Carter, HI 19, and _Psammaplysilla purpurea_ (Carter), HI 25, show similar organization. Both have a relatively dense mesohyl occupied by connective tissue fibers organized into broad fiber bundles in some cases but appearing as individual fibers in others. Cellular constituents of the mesohyl are interspersed between the predominant fibrous component (Fig. 1). Specimen HI 34, (species unidentified) differs in that the mesohyl is predominantly cellular with connective tissue generally organized into relatively fine fiber tracts or as fibers containing spicules.

The widest range of connective tissue organization and complexity was found in _Thorecta_ sp., HI 45. This sponge exhibited a well defined cortex composed of fibers embedded in a matrix. Peripherally, the density of the fibrous compon-

71

ent approached that seen in vertebrate tendons (Fig. 2).

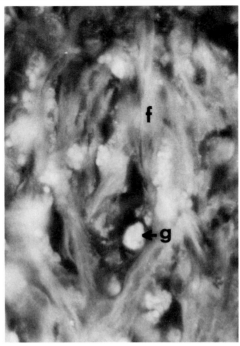

Figure 1. Prominent tracts of spongin fibers, f., surround ameboid granulocytes, g., of the mesohyl of <u>Geodia globo-stellifera</u> Carter. Fluorescein mercuric acetate for demonstration of protein sulfhydryl groups. 530 nm barrier filter.

Merging with the fibrous cortex is a zone of hyaline "chondrochyme"--using Minchin's (1900) term because of the strong resemblence to cartilaginous tissues. In the chondrochyme (Fig. 2), isolated cells are scattered within a homogenous matrix which also carries isolated fibers and bundles of spongin.

Subjacent to the cortex are found prominent, laminated masses of spongin. Although these may contain spicules, more often they are simply homogenous (but chemically heterogeneous) layers of spongin (Figs. 3 & 4).

The remainder of the sponge is the rather dense choanosome characterized by possible multi-flagellate choanocytes and a uniform matrix containing isolated fibers.

The spongillid <u>Ephydatia fluviatilis</u> is loosely organized with most spongin being associated with fasicles or individual megascleres (Fig. 5). A basal spongin layer is prominent but an organized mesohylar connective tissue is absent as such.

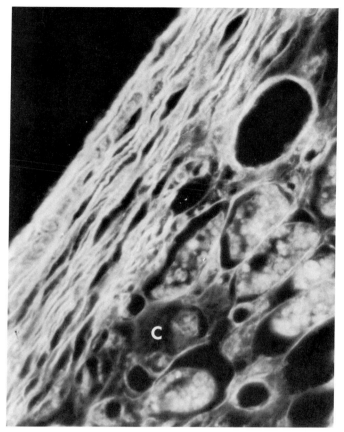

Figure 2. The fibrous cortex of <u>Thorecta</u> <u>sp</u>. demonstrates an organization pattern closely resembling that seen in vertebrate tendon. Subtending the peripheral fibrous cortex, the chondrochyme, c., shows characteristics seen in hyaline connective tissue of other groups. Fluorescein mercuric acetate. 530 nm barrier filter.

Table 1 illustrates the chemical heterogeneity seen in sponge connective tissues, even within one species. For example, in the specimen, <u>Thorecta</u> <u>sp</u>., the large spongin trabeculae contain no acid mucins while both the cortical fibers and the hyaline chondrochyme matrix contain both carboxylate and sulfomucins. Trabeculae and fibers contain considerable basic protein while the chondrochyme matrix contains little. In similar fashion, C-terminal and side chain carboxyl groups

73

are demonstrable in trabeculae but are found in very low levels in fibers and matrix areas.

Figure 3. Laminated spongin trabecula of Thorecta sp. seen in cross-section. Note distinctive localization of protein sulfhydryl fluorescence. Fluorescein mercuric acetate. 530 nm barrier filter.

The level of protein SH and SS groups present in all sponge connective tissue elements is striking. Disulfide levels appear particularly high in spongin trabeculae of Thorecta sp., but protein-bound SS groups are present in all connective tissue elements, including matrix regions. In the trabeculae, FMA procedures for SH and SH + SS reveal a distinct lamination of SH and SS distribution (Figs. 3 & 4). Regions strongly fluorescent for protein-SH contain little protein-SS and vice-versa.

DISCUSSION

The cytochemical demonstration of high levels of protein disulfides in sponge connective tissue is of considerable interest. In vertebrate collagens, cystine or cysteine have been demonstrated only in the collagen precursor, procolla-

gen, and in basement membrane collagen. The procollagen
molecule is longer than the collagen molecule in that addit-

Figure 4.   Laminated spongin trabecula from <u>Thorecta</u> <u>sp</u>., an
adjacent section to that seen in figure 3, fluorochromed for
demonstration of the sum of protein sulfhydryl plus disulfide
groups.   Note considerable increase in fluorescence output,
indicating protein disulfide localization.   Fluorescein mercuric acetate after thioglycollate reduction.   530 nm barrier
filter.

ional amino acid sequences are found at the NH_2-terminal end
of each polypeptide chain (Bellamy and Bornstein, '71; Dehm
et al., '72; Layman et al., '71).   The discovery of cysteine
in procollagen (Lenaers et al., '71) raised the possibility
that the three pro-α chains were linked together by disulfide bonds not found in collagen itself.   Schofield et al.
('74) showed that the major part of both the newly synthesized intracellular procollagen and the secreted procollagen consisted of pro-α chains linked by disulfide bonds.
These authors suggest that formation of interchain disulfide
bonds may facilitate correct association and alignment of the
three polypeptide chains to promote formation of the helical
structure of collagen.

Significant amounts of half cystine residues have been found in various basement membrane collagens (Kefalides, '72, & '73). Newly synthesized basement membrane collagen is thought to be secreted as procollagen (Grant et al., '72). Studies with basement membranes isolated from adult animals suggest that the basement membrane collagen is deposited in the extracellular space as the precursor procollagen and is not converted to the respective collagen form (Kefalides, '71).

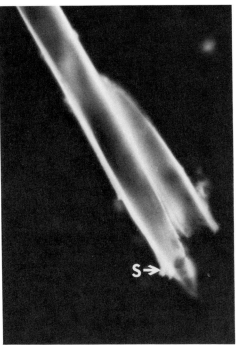

Figure 5. Megascleres of the freshwater sponge, Ephydatia fluviatilis, are bound into fasicles by a disulfide-rich spongin connective tissue sheath, s. Fluorescein mercuric acetate after thioglycollate reduction. 530 nm barrier filter.

Recent analyses of sponge collagens (Junqua et al., '74) and sponge intercellular matrix (Junqua et al., '75) have demonstrated significant levels of half-cystine residues in both. Junqua et al. ('74) suggest that sponge collagens resemble vertebrate basement membrane collagens on the basis of high hydroxylysine and hydroxylysine-bound disaccharide contents and the high 3-hydroxyproline content. Examination of analyses published by Kefalides ('73) and Junqua et al. ('74)

indicates that in terms of such criteria as percentage of total hydroxyproline as 3-hydroxyproline, number of hydroxylysine residues/1000, sum of lysine + hydroxylysine residues,
isoleucine and leucine values, and alanine values, the degree
of relationship of sponge collagens to either interstitial
collagen or basement membrane collagen is quite variable.
This should be expected in such widely separated taxa. However, examination of half-cystine values appear to show a
striking relationship to basement membrane collagen.

TABLE 1

Cytochemical Characteristics of Sponge Connective Tissue

Specimen	Masson	Neutral mucins	Sulfated mucins	Carboxylated mucins	Ionizable amino groups	Lysine	Tyrosine	Side-chain Carboxyls	C-terminal Carboxyls	Sulfhydryls (SH)	Disulfides (SS) + SH
Geodia globo-stellifera											
fibers	deep blue	3^+-4^+	2^+	3^+	3^+	2^+-3^+	1^+-2^+	$-^+$-1^+	2^+	3^+	3^+
matrix	pale blue	1^+	?	2^+	?	?	$-^+$-1^+	?	- to 1^+	1^+	2^+
granular cells	purple	—	—	—	4^+	4^+	4^+	1^+-2^+	4^+	2^+-3^+	3^+
Psammaplysilla purpurea											
fibers	deep blue	3^+-4^+	2^+	2^+-3^+	3^+-4^+	2^+	2^+-3^+	1^+-2^+	4^+	3^+-4^+	2^+
matrix	red	2^+	?	2^+-3^+	?	?	1^+	?	?	?	?
encapsulation											
fibers	?	3^+-4^+	?	?	?	?	4^+	?	4^+	?	?
granular cells	red	3^+	—	—	1^+	2^+	3^+	?	2^+	1^+	1^+
Heron Island #34											
spicular sheath	deep blue	3^+	1^+	1^+	4^+	3^+-4^+	4^+	3^+-4^+	3^+	3^+-4^+	2^+
fiber tracts	pale blue	3^+	?	?	2^+-3^+	2^+	?	2^+-3^+	2^+	3^+-4^+	2^+
Thorecta sp.											
spongin trabeculae	bright red; deep blue layers	4^+	—	—	4^+	4^+	4^+	1^+-3^{+**}	4^+	- to 4^{+**}	1^+-4^+
hyaline chondrochyme											
matrix	pale blue	2^+	3^+-4^+	3^+	- to 1^+	- to 1^+	1^+	—	1^+	1^+-2^+	2^+
fibers	deep blue	3^+	3^+-4^+	3^+-4^+	2^+-3^+	3^+-4^+	2^+	1^+	2^+	2^+-3^+	2^+-3^+
fibrous cortex											
matrix	pale blue	2^+	2^+	2^+-3^+	- to 1^+	- to 1^+	1^+	—	1^+	1^+-2^+	2^+
fibers	deep blue	3^+	3^+-4^+	3^+-4^+	2^+-3^+	2^+	2^+	1^+	2^+	3^+-4^+	3^+-4^+
choanosome											
choanocytes	reddish	2^+-3^+	2^+	2^+-3^+	3^+	3^+-4^{+*}	3^+	2^+	4^{+*}	3^+	3^{+*}
matrix	reddish	1^+-2^+	2^+	2^+-3^+	3^+	2^+-3^{+v}	2^+-3^+	2^+	2^+-3^+	3^+	3^+
granular cells	pale blue	1^+-2^+	- to 1^+	—	1^+-2^+	?	?	2^+	2^+-4^+	2^+	2^+
Ephydatia fluviatilis											
spicular sheath	pale blue	3^+-4^+	—	—	3^+-4^+	2^+	4^+	2^+-3^+	1^+-3^+	3^+-4^+	3^+-4^+
basal spongin	?	?	—	—	3^+	4^+	?	3^+	4^+	3^+	2^+
encapsulation											
fibers	?	?	—	—	4^+	4^+	?	3^+	4^+	3^+	2^+
granular cells	?	3^+	—	—	4^+	4^+	3^+-4^+	—	?	3^+	?

SYMBOLS: Range from (—) to (4^+), negative to high intensity, respectively.
The symbol (?) indicates an inability to make an observation due to absence
of representative material or masking of component.
*Nuclei.
**Layering present with variability of intensity in layers.

These data seem to suggest that the relationship is not
to basement membrane collagen per se, but rather reflect the
probable precursor state of both basement membrane collagen
and sponge collagen. It is proposed that in the evolution

of connective tissue systems, collagen evolved in the most primitive metazoans, the sponges, in a form similar to pro-collagen. The demonstration of significant levels of half-cystine residues in Metridium mesogleal collagens (Pik-karainen et al., '68), in collagen molecules isolated from Ascaris lumbricoides cuticle (McBride and Harrington, '67) and from the smooth muscle layer of Ascaris body wall (Fujimoto et al., '69) suggests a collagen bearing some similarities to procollagen may have been retained as a connective tissue component in at least some other inverte-brate phyla. The report by Pikkarainen et al. ('68) that sea anemone collagens contain three identical α-chains, a char-acteristic of some lower vertebrate collagens (Pikkarainen and Kulonen, '65) and also of basement membrane collagen (Kefalides, '73), would appear to support this hypothesis.

Leaving aside the cytochemical, biophysical, and biochem-ical characterization of sponge exoplasmic connective tissue elements, the morphological findings in themselves are wor-thy of special comment. Each group of specialists interest-ed in poriferan biology have evolved their own criteria for evaluating diversity and specialization. No zoologist who has ever examined a reef containing a variable sponge popu-lation, and who has handled these organisms, can fail to re-mark on the variability in size, consistency, shape, and friability of sponge tissues. Our investigations indicate that almost the full range of connective tissue specializa-tions encountered in the animal kingdom are met in sponges. These include: loose fine fibrous tissue; loose fine fibrous tissue in denser but irregular bundles; branching spongin main support structures with fine fibrous tissue secondary support; spongin with spicules embedded within them; tough and brittle tendon-like organizations in the sense of dense regular connective tissue; a mixture of fine fibrous con-nective tissue but with extracellular "ground substance" pre-dominating; and chondrochyme which resembles hyaline carti-lage in may ways.

It would be reasonable in the light of these findings to propose that, in addition to other considerations, evolution in sponges also represents a specialized adventure in modu-lation of connective tissue organization. The material pre-sented is far from a detailed survey; it represents more or less detailed study of some selected specimens from two lo-cations. The results have produced a sufficient insight into sponge connective tissue diversity to warrant extension of this study to other species, and other sites. Further, these results again underscore the great diversity en-countered within the groups and the need for specific detailed experimental investigation of the embryonic origin of these configurations, their regrowth in regeneration and the general cellular reaction to various types of foreign bodies and substances. In short, we are just beginning to understand possibilities and dimensions of a very large and extremely interesting problem in connective tissue biology which recognizes that the first major successful metazoan phylum has evolved the full range of modulations of connective tissue organizations encountered in higher animals. As might be expected, the best current experimental evidence suggests as well that the molecular mechanisms involved are similar.

ACKNOWLEDGMENTS

Australo-Pacific marine sponges used in this study were collected while F.W.H. was a Visiting Investigator, Heron Island Research Station, Great Barrier Reef, Queensland, Australia. These specimens were identified by Dr. Willard Hartman, Peabody Museum, Yale University, and by Dr. Patricia Bergquist, Department of Zoology, University of Aukland, New Zealand. Specimens of E. fluviatilis were collected and donated by Dr. Michael Poirrier, New Orleans University.

LITERATURE CITED

Bellamy, G. and P. Bornstein 1971 Evidence for procollagen, a biosynthetic precursor of collagen. Proc. Natl. Acad. Sci. USA, 68: 1138-1142.

Borojević, R. and P. Lévi 1967 Le basopinacoderme de l'éponge Mycale contarenii (Martens). Technique d'étude des fibres extracellulaires basales. J. microscopie, 6: 857-862.

Cowden, R.R. 1970 Connective tissue in six marine sponges: a histological and histochemical study. Zeitschr. f. mikro.-anat. Forschung., 82: 557-569.

Cowden, R.R. and S.K. Curtis 1970 Demonstration of protein-bound sulfhydryl and disulfide groups with fluorescent mercurials. Histochemie, 22: 247-255.

Curtis, S.K. and R.R. Cowden 1971 Fluorescence of 2-hydroxy-3-naphthoic acid hydrazide derivatives of side-chain groups of proteins. Histochemie, 28: 345-350.

Dehm, P., S.A. Jimenez, B.R. Olsen, and D.J. Prockop 1972 A transport form of collagen from embryonic tendon: electron microscopic demonstration of an NH_2-terminal extension and evidence suggesting the presence of cystine in the molecule. Proc. Natl. Acad. Sci. USA, 69: 60–64.

Deitch, A.C. 1955 Microspectrophotometric study of the binding of the anionic dye, naphthol yellow S by tissue sections and by purified proteins. Lab Invest., 4: 324–351.

DeVos, L. 1972 Fibres géantes de collagène chez l'éponge Ephydatia fluviatilis. J. Microscopie, 15: 247–252.

DeVos, L. and F. Rozenfeld 1974 Ultrastructure de la coque collagène des gemmules d'Ephydatia fluviatilis (Spongilli-des). J. Microscopie, 20: 15–20.

Fujimoto, D., T. Ikeuchi, and S. Nozawa 1969 Intermolecular disulfide cross-linkages in the collagen from the muscle layer of Ascaris lumbricoides. Biochem. Biophys. Acta, 188: 295–301.

Garrone, R. 1969 Collagène, spongine et squellette minéral chez l'éponge Haliclona rosea (O.S.) (Démosponge, Haplo-scléride). J. Microscopie, 8: 581–598.

Garrone, R. and M. Pavans de Ceccatty 1971 Histophysiologie ultrastructurale du tissue conjonctif. Lyon med., 226: 725–735.

Garrone, R., J. Vacelet, M. Pavans de Ceccatty, S. Junqua, L. Robert and A. Huc 1973 Une formation collagène par-ticulière: les filaments des éponges cornées Ircinia. Étude ultrastructurale, physico-chemique et biochimique. J. Microscopie, 17: 241–260.

Grant, M.E., N.A. Kefalides and D.J. Prockop 1972 The biosynthesis of basement membrane collagen in embryonic chick lens. Delay between the synthesis of polypeptide chains and the secretion of collagen by matrix-free cells. J. Biol. Chem., 247: 3539–3544.

Gross, J., Z. Sokal and M. Rougvie 1956 Structural and chemical studies on the connective tissue of marine sponges. J. Histochem. Cytochem., 4: 227–246.

Harrison, F.W., S.K. Curtis, and R.R. Cowden 1975 Evalua-tion of two protein end-group reactions as potential fluo-rescent cytochemical methods. Histochem. Jour., 7: 91–94.

Huc, A. and L. de Vos 1972 La nature collagène de la coque des gemmules des éponges d'eau douce. C. r. hebd. Seanc. Acad. Sci., Paris, 275: 1399–1401.

Junqua, S., J. Fayolle and L. Robert 1975 Structural glyco-proteins from sponge intercellular matrix. Comp. Biochem. Physiol., 50B: 305–309.

Junqua, S., L. Robert, R. Garrone, M. Pavans de Ceccatty and J. Vacelet 1974 Biochemical and morphological studies on collagens of horny sponges. Ircinia filaments compared to spongines. Conn. Tiss. Res., 2: 193–203.

Kefalides, N.A. 1971 Chemical properties of basement mem-

branes. Int. Rev. Exp. Pathol., 10: 1-39.
_____ 1972 The chemistry of antigenic components isolated
from glomerular basement membrane. Conn. Tissue Res.,
1: 3-13.
_____ 1973 Structure and biosynthesis of basement membranes.
Int. Rev. Conn. Tissue Res., 6: 63-104.
Kefalides, N.A. and B. Denduchis 1969 Structural components of epithelial and endothelial basement membranes.
Biochemistry, 8: 4613-4621.
Layman, D.L., E.B. McGoodwin, and G.R. Martin 1971 The
nature of the collagen synthesized by cultured human fibroblasts. Proc. Natl. Acad. Sci. USA, 68: 454-458.
Ledger, P.W. 1974 Types of collagen fibers in the calcareous sponges Sycon and Leucandra. Tissue Cell, 6: 385-389.
Leeman, U. and F. Ruch 1972 Cytofluorometric determination
of basic and total proteins with sulfaflavine. J. Histochem. Cytochem., 20: 659-671.
Lenaers, A., M. Ansay, B.V. Nusgens, and C.M. Lapiere 1971
Collagen made of extended αchains, procollagen, in ge-
netically-defective dermatosparaxic claves. Eur. J. Biochem., 23: 533-543.
Marks, M.H., R.S. Bear, and C.H. Blake 1949 X-ray diffrac-
tion evidence of collagen-type protein fibers in the
Echinodermata, Coelenterata, and Porifera. J. Exp. Zool.,
111: 55-78.
McBride, O.W. and W.F. Harrington 1967 Ascaris cuticle collagen: on the disulfide cross-linkages and the molecular
properties of the subunits. Biochem., 6: 1484-1498.
Minchin, E.A. 1900 Porifera. In Lankester, E.R., ed.,
Treatise on Zoology. Black, London, pp. 1-178.
Mowry, R.W. 1963 The special value of methods that color
both acidic and vicinal hydroxyl groups in the histochem-
ical study of mucins with revised idrections for the col-
loidal iron stain, the use of alcian blue 8 GX and their
combination with the Periodic acid-Schiff reaction. Ann.
N.Y. Acad. Sci., 106: 402-423.
Pavans de Ceccatty, M. and Y. Thiney 1963 Microscopie élec-
tronique de la fibrogenèse cellulaire du collagène, chez
l'éponge siliceuse Tethya lyncurium LK, C. r. hebd. Seanc.
Acad. Sci., Paris, 256: 5406-5408.
Pikkarainen, J. and E. Kulonen 1965 Collagen of lamprey.
Acta Chem. Scand., 19: 280.
Pikkarainen, J., J. Rantanen, M. Vastamäki, K. Lampiaho, A.
Kari and E. Kulonen 1968 On collagens of invertebrates
with special reference to Mytilus edulis. Europ. J. Bio-
chem., 4: 555-560.
Rigler, R. Jr. 1966 Microfluorometric characterization of

intracellular nucleic acids and nucleoproteins by acridine orange. Acta Physiol. Scand., 67: Suppl. 267: 1-122.

Ritter, C. and J. Berman 1963 The quantitative spectro-photometric analysis of tyrosine by a modified diazoti-zation-coupling method. J. Histochem. Cytochem., 11: 590-602.

Rosselet, A. and F. Ruch 1968 Cytofluorometric determina-tion of lysine with dansylchloride. J. Histochem. Cyto-chem., 16: 459-466.

Schofield, J.D., J. Uitto, and D.J. Prockop 1974 Forma-tion of interchain disulfide bonds and helical structure during biosynthesis of procollagen by embryonic tendon cells. Biochemistry, 13: 1801-1806.

Simpson, T.L. 1968 The structure and function of sponge cells: New criteria for the taxonomy of poecilosclerid sponges (Demospongiae). Bull. Peabody Mus. Nat. Hist., 25: 1-142.

Stoward, P.J. and J. Burns 1967 Studies in fluorescence histochemistry. IV. The demonstration of the C-terminal carboxyl groups of proteins. Histochemie, 10: 230-233.

Swift, H. 1966 The quantitative cytochemistry of RNA, In Weid, G.L., ed., An Introduction to Quantitative Cyto-chemistry. Academic Press, New York, pp. 255-286.

Vacelet, J. 1971 Ultrastructure et formation des fibres de spongine d'éponges cornées Verongia. J. Microscopie, 10: 13-32.

RECENT INVESTIGATIONS OF THE INVOLVEMENT OF 3', 5' CYCLIC AMP IN THE DEVELOPMENTAL PHYSIOLOGY OF SPONGE GEMMULES[1]

Tracy L. Simpson and Gideon A. Rodan

Department of Biology, University of Hartford, West
Hartford, Connecticut 06117 and School of Medicine
and Dental Medicine, University of Connecticut,
Farmington, Connecticut 06032

ABSTRACT: Dormant gemmules of the fresh-water sponge,
Spongilla lacustris at $3^{\circ}C$ storage contain 2 pmoles of
cyclic AMP per gemmule. Upon germination at $20^{\circ}C$ the
cyclic AMP rapidly drops to 0.8 pmoles/gemmule. Incubation of gemmules in 7 mM theophylline, an inhibitor
of phosphodiesterase, prevents germination, keeping
the thesocytes in their original binucleate condition.
At lower concentrations of theophylline, germination
(cell division) and hatching (cell attachment and
spreading) occur and are rapidly followed by formation
of a new gemmule, a response never seen in controls.
Exogenous cyclic AMP, N-but cyclic AMP, epinephrine,
and cyclic GMP also cause new gemmule formation. Crude
preparations of gemmules possess adenylate cyclase,
guanylate cyclase, and phosphodiesterase activity.
These findings support the view that cyclic nucleotides
and the activity of the corresponding enzymes are intimately involved in the control of dormancy and other
developmental stages in the life cycle of these animals. Gemmulostasine (a diffusible substance derived
from newly hatched sponges) and increases in osmotic
pressure also inhibit germination. We suggest in both
of these that changes in the levels of cyclic AMP are
produced possibly through partial inhibition of phosphodiesterase activity.

During the past year we have initiated investigations to
determine the possible involvement of adenosine 3', 5'
cyclic monophosphate (cAMP) in gemmule germination of the
freshwater sponge Spongilla lacustris. Two papers have appeared recently from the laboratory of Professor R. Rasmont
in Bruxelles describing laboratory stimulation of gemmule
formation by theophylline in Ephydatia fluviatilis (Rasmont,
'74; Rasmont and DeVos, '74). In view of this recent work,

[1]This work was supported by NSF Grant GB-37775 and a grant
from the University of Connecticut Research Foundation.

summarized below and the results presented here it now appears that cAMP, adenylate cyclase, and cAMP phosphodiesterase (PDE) play an important role in gemmule dormancy, germination, and gemmule formation. The involvement of cyclic nucleotides was best documented in gemmule germination through cAMP radioimmunoassays and measurements of adenylate cyclase and phosphodiesterase activities. The process of gemmule germination and formation is described in detail in Simpson and Fell ('74).

MATERIALS AND METHODS

Gemmules: Gemmules of the fresh-water sponge Spongilla lacustris were collected in early November from a bog pond in New Hampshire and stored in the dark at 4°C. In all experiments, gemmules were dissected from the parent skeleton at 4°C in a pond water simulated medium (Medium M) (Rasmont, '61) and groups were matched for gemmule size.

Nuclear Counts: Incubated and unincubated gemmules were fixed, embedded in epon as previously described (Simpson and Vaccaro, '74), and 1μm toluidin-blue-stained sections were examined at a magnification of 1000x. Five fields were selected at random from sections 40μm apart.

cAMP and cGMP Assays: Gemmules dissected from the same sponge were matched for size and divided into groups of 8-10. Each group was incubated in 10ml of Medium M with or without aminophylline (theophylline$_2$ ethylene diamine). For each condition and at each time point, cAMP or cGMP were determined in 3-8 groups of gemmules by a modification of a previously reported method (Rodan et al., '75). All 8-10 gemmules of a group were homogenized at 4°C in 0.5ml Krebs-Ringer-Tris buffer (NaCl 120mM, KCl 4.8mM, $CaCl_2$ 1.2mM, $MgSO_4$ 1.2mM, Tris-HCl and Tris-Base 15mM, glucose 2gm/l, pH 7.4) in a small ground glass tissue homogenizer (VWR). The homogenate was boiled for five minutes, sonicated for 20 seconds at 50 watts (Sonifer Cell Disrupter, Model W185, Heat Systems-Ultrasonics, Inc., Plainview, N.Y.) and centrifuged at 5000g for 10 minutes at 4°C. cAMP or cGMP were determined in duplicate 100μl aliquots of the supernatant by radioimmunoassay (Steiner, et al., '69) (Collaborative Research, Waltham, Mass.). The recovery rate of cAMP was 86-94% as determined by addition of [3]HcAMP at the beginning of the preparative procedure.

Adenylate and Guanyl Cyclase Assays: Twenty gemmules were homogenized at 4°C in a small ground glass tissue homogenizer (VWR) in 100μl of 0.25 M sucrose, 0.002 M $MgCl_2$, 0.01 M Tris, pH 7.4. Adenylate cyclase activity was measured by the meth-

od of Salomon et al. ('75) and guanyl cyclase by the method
of White and Zenser ('71) in an aliquot of the homogenate in
a volume of 100µl assay mixture of the following composition:
25mM Tris HCl pH 7.6, 5mM $MgCl_2$, 1mM cAMP or cGMP, 1mM di-
thiothreitol (DTT), 0.1mM ATP or GTP, 3 x 10^6 cpm α-^{32}P-ATP
or α-^{32}P-GTP, 10mM phosphocreatine, 10 units phosphocreatine
kinase, approximately 100µg protein as determined by the
method of Lowry et al. ('51) using bovine serum albumin as
standard.

<u>cAMP and cGMP Phosphodiesterase Assays</u>: Homogenates were
prepared as described above for the cyclase assays. cAMP and
cGMP phosphodiesterase (PDE) activities were measured in
100µl assay mixture of the following composition: 80mM
imidazole pH 6.9, 0.3mM DTT, 3.4mM $MgCl_2$, 6.7mM 5'AMP or
5'GMP, 0.01mM cAMP or cGMP, 10^5 cmp ^{3}HcAMP or ^{3}HcGMP, final
pH 7.3. The reaction was stopped by boiling. The reaction
products ^{3}H-5' AMP and ^{3}H-5' GMP were separated from ^{3}HcAMP
and ^{3}HcGMP on short, acidified Ag50WX4 200-400 mesh (Bio
Rad) columns. 2 x 10^3 cpm ^{14}C-5' AMP or ^{14}C-5' GMP in 10µl
were added before chromatography to determine product recovery.

For cAMP PDE one ml of a 25% suspension (w/v) of washed
Ag50WX4 resin was pipetted into pasteur pipettes stoppered
with glass wool. Following one ml of 1 N HCl and three one
ml water washes the reaction mixture was layered on the
resin. The first 4 ml of water eluate was discarded and the
second 4 ml containing 40-50% of the ^{3}H-5' AMP was collected
and counted in 10ml of Bray scintillation fluid. For cGMP
PDE three ml of the 25% resin was used to make the columns.
The first 2ml of water eluate was discarded and the second
2ml containing 60-70% of the ^{3}H-5' GMP was collected in 10ml
of Bray. A boiled enzyme blank was run in all experiments.
Samples were counted in an Isocap 300 (Nuclear Chicago) using
the double label channel for ^{3}H and ^{14}C. Counts were cor-
rected by substraction of blank values, for overlap, and recovery.

RESULTS

Cyclic Nucleotides in Cold Stored Gemmules
   We have examined cold stored gemmules of <u>S. lacustris</u> by
radioimmunoassay and have identified both cAMP and guanosine
3', 5' cyclic monophosphate (cGMP) in them. There is approximately 10-20 times more cAMP present than cGMP (Table 1).
We have furthermore identified the corresponding cyclic nucleotide enzymes in these gemmules: these include low and
high affinity cAMP PDE, cGMP PDE, adenylate cyclase, and

85

TABLE 1

CYCLIC NUCLEOTIDE CONTENT AND ACTIVITY OF CYCLIC NUCLEOTIDE
ENZYMES IN DORMANT GEMMULES OF SPONGILLA LACUSTRIS

Cyclic Nucleotide	pmoles/gemmule
cAMP	2.14 ± 0.22[1]
cGMP	0.18 ± 0.08[2]

Enzyme	Activity (pmoles/min/mg protein)[3]
Guanyl cyclase	Present
Adenylate cyclase	2.0
Low affinity cAMP PDE[4]	1,677.0
High affinity cAMP PDE[5]	99.0
cGMP PDE[5]	120.0

[1]Mean and SEM of measurements of 82 gemmules.
[2]Mean and SEM of measurements of 30 gemmules.
[3]Measurements at 19-21°C. For each enzyme measurements were
made in duplicate at five minute intervals during 30 minute
incubation of polled homogenates of cold stored gemmules.
Activities listed were determined by linear regression.
[4]Substrate concentration, 650μM.
[5]Substrate concentration, 10μM.

guanyl cyclase; the latter has only been measured at one time
point on two occasions. In our work thus far we have found
considerable variability in basal activities from one preparation to another which may be a reflection of differences in
the crude homogenates employed for the measurements. Ideal-
ly, partially purified preparations should be employed but
this requires relatively large quantities of gemmules. The
activity of the low affinity cAMP PDE may be partially due to
a non-specific esterase identified histochemically in gemmules of Ephydatia fluviatilis by Tessenow ('69). Based upon
the high substrate concentration required to study its activity, the low affinity cAMP PDE is of little physiological
significance relative to cyclic nucleotide metabolism; additional studies, however, are needed. The high and low affinity cAMP PDEs have different temperature dependencies: the
high affinity displays maximum activity at 15-20°C and minimum activity at 37°C; the low affinity enzyme is roughly
linear throughout the temperature range of 4-37°C with maximum activity at 37°C (Fig. 1). Adenylate cyclase activity
increases up to 25°C and then levels off. The activity of
the high affinity cAMP PDE is not significantly affected by
addition of equimolar cGMP; cGMP PDE is likewise unaffected

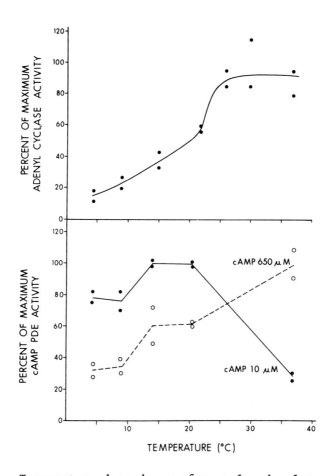

Fig. 1. Temperature dependency of gemmule adenylate cyclase (above) and cAMP phosphodiesterase (below) activities. For each enzyme the activity was assayed in a pooled homogenate of cold stored gemmules. Two substrate concentrations used for cAMP PDE were: 10μM (●-●) for the high affinity enzyme and 650μM (o--o) for the low affinity enzyme. Results of two measurements at each temperature are expressed as percent of maximum activity, which was 0.2 pmoles/min/mg protein for adenylate cyclase, 97.0 pmoles/min/mg protein for the high affinity cAMP PDE and 2,866.0 pmoles/min/mg protein for the low affinity cAMP PDE. The low affinity PDE while displaying linearity tends to flatten at 15-20°C; this may be due to a contribution of the high affinity enzyme at this concentration.

by cAMP (Table 2). These results indicate that we are dealing with two distinct enzymes with little or no physiological overlap in substrate specificity.

TABLE 2

EFFECT OF cAMP AND cGMP ON THE ACTIVITY OF cAMP PDE AND cGMP PDE IN DORMANT GEMMULES OF <u>SPONGILLA LACUSTRIS</u>

Enzyme	Activity (pmoles/min/mg protein)[1]	
	10 μM cAMP Substrate	10 μM cAMP + 10 μM cGMP Substrate
cAMP PDE	159.4 ± 1.1	144.7 ± 4.6
	10 μM cGMP Substrate	10 μM cGMP + 10 μM cAMP Substrate
cGMP PDE	179.5 ± 6.3	192.6 ± 12.8

[1]Duplicate measurements at $20^{\circ}C$ on pooled homogenates of cold stored gemmules following 15 minute incubation. Means and SEM.

Changes in cAMP, Adenylate Cyclase and cAMP PDE during Gemmule Germination

In the first few hours of incubation at $20^{\circ}C$ the cAMP content of germinating gemmules falls from about 2 pmoles/gemmule to less than 1 pmole/gemmule (Fig. 2). During this period there is a substantial reduction in the activity (at $20^{\circ}C$) of adenylate cyclase but little change in cAMP PDE (Table 3). However, since the cAMP PDE in cold stored gemmules shows an increase in activity at $20^{\circ}C$ as compared to $4^{\circ}C$ (Fig. 1) this enzyme is more active during germination than during dormancy.

Effect of Aminophylline on Gemmule Germination, cAMP and cAMP PDE

When gemmules of S. lacustris are incubated in the presence of $7 \times 10^{-3}M$ aminophylline (theophylline$_2$ethylene diamine) they fail to hatch. This is due to the arrest of cell division (Table 4) and probably also of nuclear separation, the earliest observable morphological event during germination (Berthold, '69). $7 \times 10^{-4}M$ aminophylline also inhibits hatching; however, the decrease in binucleate cells indicates that either some nuclear separation and/or cell division has occurred. At 7×10^{-5} and $7 \times 10^{-6}M$ aminophylline hatching occurs and thus cell division is not affected.

88

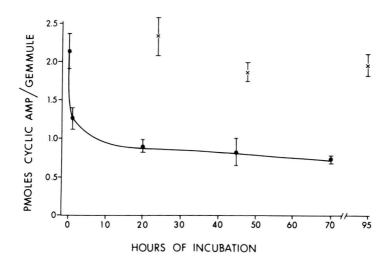

HOURS OF INCUBATION

Fig. 2.   Changes in cAMP content during incubation at 20°C in the absence (•-•) or presence (x) of 7mM aminophylline. The results are means and standard errors of the means of 3-8 groups (see Materials and Methods) of gemmules for each point.   In the absence of aminophylline hatching occurs at 70 hours.

TABLE 3

CHANGES IN ADENYLATE CYCLASE AND cAMP PHOSPHODIESTERASE ACTIVITIES DURING EARLY GERMINATION[1]

	Adenylate Cyclase (pmoles/min/mg protein)	cAMP Phosphodiesterase (pmoles/min/mg protein)
Cold stored gemmules	0.27 ± 0.01	65.9 ± 0.3
After two hour incubation at 20°C	0.16 ± 0.001	56.4 ± 3.8

[1]All gemmules were dissected at 4°C from the same sponge and were divided into groups of 20, matched for gemmule size. Half the groups were kept at 4°C and half at 20°C. At the end of two hours each group was assayed for enzyme activity. The results are means and standard errors of the means of four determinations each. Incubations were for 20 minutes at 20°C. For cAMP PDE the substrate concentration was 10 µM cAMP.

TABLE 4

EFFECTS OF AMINOPHYLLINE ON CELL DIVISION AND
HATCHING IN GEMMULES OF SPONGILLA LACUSTRIS

Treatment	Percent binucleate[1] cells at 48 h	Hatching at 5 days
Cold stored gemmules (3°C)	64	−
Control (20°C)	6	+
Aminophylline (20°C)		
7 X 10^{-6} M	Unrecorded	+
7 X 10^{-5} M	Unrecorded	+
7 X 10^{-4} M	36	−
7 X 10^{-3} M	54	−

[1]Percent of total nucleated cells counted.

Measurement of the cAMP content of gemmules incubated at 20°C
in 7 x 10^{-3} M aminophylline shows that this concentration also inhibits the reduction in cAMP content during the period
of germination (Fig. 2). The in vitro activity of cAMP PDE
in cold stored gemmules is inhibited 70-90% by 7 x 10^{-4} M
and higher concentrations of aminophylline; 7 x 10^{-5} and
7 x 10^{-6} produce little (0-20%) inhibition of PDE (Fig. 3).
Effect of Cyclic Nucleotides, Theophylline, and Aminophylline
on Newly Hatched Gemmules

Two types of experiments have been performed in which newly
hatched sponges were exposed to exogenous cyclic nucleotides,
theophylline, and aminophylline. In the case of the cyclic
nucleotides different results have been obtained. Rasmont
('74) has added cyclic nucleotides and theophylline to seven
day old sponges of Ephydatia fluviatilis, each sponge being
formed from the hatching of four gemmules. Concentrations of
4 x 10^{-4} to 4 x 10^{-5} M theophylline stimulate the formation
of new gemmules by all sponges in four to nine days. In this
same system cAMP, N^6, $O^{2'}$-dibutyryl adenosine 3', 5' cyclic
monophosphate (dibutyryl cAMP), and cGMP from 10^{-9} to 10^{-3} M
did not cause the formation of new gemmules. Control sponges
in these experiments also did not form gemmules during the
experimental period.

We have carried out similar experiments with Spongilla
lacustris in which sponges hatched from single gemmules were
placed in the solutions at the beginning of germination. The
results are presented in Table 5. Figures 4 and 5 are photo-

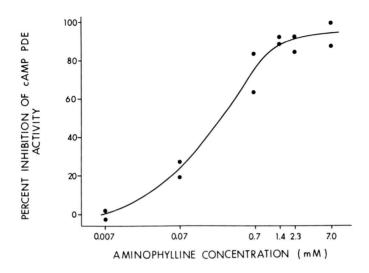

Fig. 3. Effect of aminophylline on the activity of gemmule cAMP phosphodiesterase. Enzyme activity was assayed in a pooled homogenate of cold stored gemmules at the indicated aminophylline concentrations. Results of duplicate measurements are expressed as percent inhibition relative to the maximum activity (51 pmoles/min/mg protein) observed in the absence of aminophylline. The assay mixture contained 10 μM cAMP. The reaction was run for 15 minutes at 20°C.

graphs of sponges continuously exposed to 7×10^{-5}M aminophylline and illustrate its stimulation of cell aggregation and gemmule formation.

DISCUSSION

These results show that a drop in cAMP content and an active cAMP PDE are necessary for the germination of gemmules of S. lacustris to occur. The following data lead to this conclusion: (i) in germinating gemmules a two-fold decrease in cAMP content occurs early during germination; (ii) 90% inhibition of cAMP by 7×10^{-3} M aminophylline results in the maintenance of dormant levels of cAMP and in the inhibition of cell division and probably nuclear separation; (iii) the essentiality of cAMP PDE for germination is also supported by the fact that lower concentrations of aminophylline (7×10^{-5} and 7×10^{-6} M) which produce only 0-20% cAMP PDE inhibition are not inhibitory to

91

TABLE 5

EFFECT OF CONTINUOUS TREATMENT OF SPONGILLA LACUSTRIS
GEMMULES WITH CYCLIC NUCLEOTIDES AND AMINOPHYLLINE

Treatment		5 days[1]	12 days[1]
cAMP, dibutyryl-cAMP, cGMP	10^{-5}M	Hatched	One new gemmule per sponge
Aminophylline	7 X 10^{-5}M	Hatched	One new gemmule per sponge
Aminophylline	7 X 10^{-6}M	Hatched	One new gemmule per sponge
Control		Hatched	Hatched, no gemmules

[1]Time of incubation of gemmules at 20°C. The incubation
medium was Medium M (Rasmont, '61) to which the test substances were added at time zero. Cultures consisted of five
gemmules in 10 ml of medium which was unchanged during the
experimental period. Triplicate dishes in moist chambers
were run in each case.

Fig. 4. Nine day old sponge of S. lacustris germinated and
hatched from a single gemmule, in 7 x 10^{-5} M aminophylline.
Dark areas of cell aggregation are obvious. The empty gemmule coat is also visible. Magnification approximately x 10.

germination (Fig. 3, Table 4); (iv) the role of PDE in germination is further supported by the overlap of the optimal
temperature for PDE activity (15-20°C) (Fig. 1) and that for
germination (16-20°C) as established by Rasmont ('54).

Fig. 5. Twelve day old sponge of S. lacustris germinated and hatched in 7×10^{-5} M aminophylline. The cells of the young sponge have completely aggregated to form a new gemmule. Also present is the empty, original gemmule coat and the skeleton (fibers of collagen and spicules) of the young sponge. Magnification approximately X 10.

The requirement for cAMP reduction in the process of germination is probably related to the control of cell division which is the earliest morphologically observable event in the release from dormancy. Reduced levels of cAMP have been shown in other systems to correlate with cell proliferation (Burger et al., '72; Macintyre et al., '72; Burk, '68; Johnson et al., '71; Anderson et al., '73; Sheppard, '71). In one of these which is a dormant plant system, tubers of the Jerusalem artichoke, a drop in cAMP has been described during germination (Giannattasio et al., '74). Similarly, in zoospores of Blastocladiella, a drop in cAMP was observed upon germination (Silverman and Epstein, '75).

The mechanism for the reduction in gemmule cAMP during laboratory (temperature induced) germination appears to involve both cAMP PDE and adenylate cyclase. cAMP PDE activity increased about 20% by warming the gemmules to 20°C (Fig. 1) while adenylate cyclase activity decreases approximately 40% during the first two hours of incubation at room temperature (Table 3). Gemmule germination in the laboratory may thus be the result of a new balance of activity of these enzymes caused by the increase in temperature. In the natural habitat, however, germination and hatching can occur at 4-8°C (Simpson and Gilbert, '73). It seems likely that factors other than temperature are involved. The removal of an inhibitor of PDE may control germination in the absence of warming. This hypothesis is consistent with the observation that the chemically unidentified substance gemmulostasin (Rasmont, '65) which is derived from young hatched sponges inhibits gemmule germination. Rosenfeld ('74) has shown that gemmulostasin inhibits thymidine incorporation into gemmules, which according to our findings is a cAMP like effect.

The role of cGMP, cGMP PDE, and guanyl cyclase in gemmule germination is yet to be established. In <u>Blastocladiella</u> cGMP levels increase substantially during sporulation (Silverman and Epstein, '75).

A second aspect of the involvement of cyclic nucleotides in the control of the sponge life cycle is related to gemmule formation. We found that continuous exposure of <u>S</u>. <u>lacustris</u> hatching sponges to exogenuous cAMP, cGMP or aminophylline led to gemmule formation. Rasmont produced the same effect by exposure to theophylline at concentrations similar to those effective in our system.

Critical measurements of the cAMP level of newly hatched sponges exposed to theophylline and aminophylline have to be made. But from the effects of aminophylline on the cAMP content of germinating gemmules, and its effect upon gemmule cAMP PDE it can be inferred that theophylline stimulates gemmule formation through acting upon the cyclic nucleotide metabolism of young sponges. The failure of cyclic nucleotides to stimulate gemmule formation in young sponges of <u>E</u>. <u>fluviatilis</u> (Rasmont, '74) could be due to a number of differences between that system and ours, prominent among which are the following: 1) species difference, our species contains symbiotic algae (Simpson and Fell, '74); 2) the size of the young sponges is vastly different, our sponges are four times smaller which may enhance the ability of the animal to respond; 3) the ratio of tissue to volume of medium is much lower in our experiments which reduces possible effects of inhibitors or cyclic nucleotide breakdown; and 4) we exposed the sponges to nucleotides throughout hatching. This last difference may be of significance but additional experiments are needed to clarify if it is due to the timing of exposure in the developmental sequence, differences in permeability or other factors. The mode of action of exogenous cyclic nucleotides in stimulating gemmule formation is not known but two distinct possibilities present themselves: 1) they may act only on the epithelia of the animal or 2) they may also increase the mesenchymal levels of cyclic nucleotides through diffusion. In view of the stimulation of gemmule formation by theophylline and the apparent necessity of continuous (vs pulsed) exposure to the cyclic nucleotides, the second possibility seems more likely although both may be involved.

In a series of experiments dealing with the effects of nutrition and size upon the ability of newly hatched sponges to form gemmules, Rasmont ('63) has established, in <u>E</u>. <u>fluvi-</u>

atilis, that young sponges hatched from a single gemmule fail to form a new gemmule. One of his conclusions was that a critical number of cells is necessary for gemmule formation. In S. lacustris we have also found that sponges derived from a single gemmule do not form a new gemmule. However, single-gemmule-sponges form new gemmules in the presence of cyclic nucleotides and aminophylline which demonstrates that the physiological activities of the cells is crucial rather than their number.

The specific role of cAMP and possibly cGMP in the development of these animals are questions of general significance and interest. A major question revolves around the coupling of the reduction in cAMP to the release from dormancy and its relation to cell division. Giannattasio et al. ('74) have suggested that cAMP acts as a "brake" in dormant artichoke tubers. When gemmules are considered in this context it appears that this "braking" effect is due to cyclic nucleotide inhibition of cell division.

This study establishes that sponge gemmules can be employed as a model system for further investigation of the involvement of cyclic nucleotides in development. They have the advantage of being a native system which does not require subculturing and in which the following biological phenomena, in addition to cell division and dormancy, can be investigated: cell migration, cell attachment, collagen biosynthesis, differentiation,and cell aggregation.

LITERATURE CITED

Anderson, W.B., T.R. Russell, R.A. Carchman and I. Pastan 1973 Interrelationship between adenylate cyclase activity, adenosine 3':5' cyclic monophosphate phosphodiesterase activity, adenosine 3':5' cyclic monophosphate levels, and growth of cells in culture. Proc. Nat. Acad. Sci. U.S.A., 70: 3802-3805.

Berthold, G. 1969 Untersuchungen über die histoblasten differenzierung in der gemmula von Ephydatia fluviatilis. Z. Wiss. Mikros.,69: 227-243.

Burger, M.M., B.M. Bombik, B.M. Breckenridge and J.R. Sheppard 1972 Growth control and cyclic alterations of cyclic AMP in the cell cycle. Nature, New Biology, 239: 161-163.

Giannattasio, M., E. Mandata and V. Macchia 1974 Content of 3':5' cyclic AMP and cyclic AMP phosphodiesterase in dormant and activated tissues of Jerusalem artichoke tubers. Biochem. Biophys. Res. Comm., 57: 365-371.

Johnson, G.S., R.M. Friedman and I. Pastan 1971 Restoration of several morphological characteristics of normal fibroblasts in sarcoma cells treated with adenosine 3':5' cyclic monophosphate and its derivatives. Proc. Nat. Acad. Sci. U.S.A., 68: 425-429.

Lowry, O.H., N.J. Rosenbrough, A.L. Farr and R.J. Randall 1951 Protein measurement with the folin phenol reagent. J. Biol. Chem., 193: 265-275.

Macintyre, E.H., C.J. Wintersgill, J.D. Perkins and A.E. Vatter 1972 The responses in culture of human tumor astrocytes and neuroblasts to N^6, $O^{2'}$-dibutyryl adenosine 3':5' monophosphoric acid. J. Cell Sci., 11: 639-667.

Rasmont, R. 1954 La diapause chez les spongillides. Bull. Acad. Roy. Belg., 40: 288-304.

_____1961 Une technique de culture des éponges d'eau en milieu controlé. Ann. Soc. Roy. Zool. Belg., 91: 147-156.

_____1963 La role de la taille et de la nutrition dans la determinisme de la gemmulation chez les spongillides. Develop. Biol., 8: 243-271.

_____1965 Existence d'une regulation biochimique de l'éclosion des gemmules chez les spongillides. Comp. and Rend. Acad. Sci., 261: 845-847.

_____1974 Stimulation of cell aggregation by theophylline in the asexual reproduction of fresh-water sponges (Ephydatia fluviatilis). Experientia, 30: 792-794.

Rasmont, R. and L. DeVos 1974 Étude cinématographique de la gemmulation d'une éponge d'eau douce: Ephydatia fluviatilis. Arch. Biol. (Bruxelles),85: 329-341.

Rodan, G.A., A. Harvey, T. Mensi and L.A. Bourret 1975 3':5' cyclic AMP, a possible mediator of the effects of pressure on bone remodeling. Science, 189: 467-469.

Rosenfeld, F. 1974 Biochemical control of fresh-water sponge development: Effect on DNA, RNA and protein synthesis of an inhibitor secreted by the sponge. J. Embryol. Exp. Morph., 32: 287-295.

Salomon, V., C. Londos and M. Rodbell 1974 A highly sensitive adenylate cyclase assay. Anal. Biochem., 58: 541-548.

Sheppard, J.R. 1971 Restoration of contact-inhibited growth to transformed cells by dibutyryl adenosine 3':5' cyclic monophosphate. Proc. Nat. Acad. Sci. U.S.A., 68: 1316-1320.

Silverman, P.M. and P.M. Epstein 1975 Cyclic nucleotide metabolism coupled to cytodifferentiation of Blastocladiella emersonic. Proc. Nat. Acad. Sci. U.S.A., 72: 442-446.

Simpson, T.L. and J.J. Gilbert 1973 Gemmulation, gemmule hatching and sexual reproduction in fresh-water sponges. I. The life cycle of Spongilla lacustris and Tubella pennsylvanica. Trans. Amer. Micros. Soc., 92: 422–433.

Simpson, T.L. and P.E. Fell 1974 Dormancy among the porifera: gemmule formation and germination in fresh-water and marine sponges. Trans. Amer. Micros. Soc., 93: 544–577.

Simpson, T.L. and C.A. Vaccaro 1974 An ultrastructural study of silica deposition in the fresh-water sponge Spongilla lacustris. J. Ultrastruct. Res., 47: 296–309.

Steiner, A.L., D.M. Kipnis, R. Utiger and C.W. Parker 1969 Radioimmunoassay for the measurement of adenosine 3', 5' cyclic monophosphate. Proc. Nat. Acad. Sci. U.S.A., 64: 367–373.

Tessenow, W. 1969 Lytic processes in development of freshwater sponges. In: Lysosomes in Biology and Pathology, Vol. 1, (J.T. Dingle and H.B. Fell, eds.) North Holland, Amsterdam: 392–405.

White, A.A. and T.V. Zenser 1971 Separation of cyclic 3', 5' nucleotide monophosphates from other nucleotides on aluminum oxide columns. Application to the assay of adenyl cyclase and guanyl cyclase. Anal. Biochem., 41: 372–396.

NATURAL GAMETE RELEASE AND OVIPARITY
IN CARIBBEAN DEMOSPONGIAE

Henry M. Reiswig

Redpath Museum and Department of Biology
McGill University
P.O. Box 6070
Montreal, Quebec, Canada

ABSTRACT: Reports of natural gamete release by Caribbean sponges are reviewed to determine if such events
are correlated to physical cycles. Release of spermatozoa is not seasonally limited, but it is diurnally
concentrated in the afternoon, 1400 to 1700 hours local
time. Oviparity is reported in four species among the
genera Hemectyon and Agelas (Demospongiae). Two of
these species release gametes annually at specific
seasons and stages of the lunar phase. These release
events may prove to be precisely predictable over extensive geographic ranges. Review of biochemical and
embryological evidence suggests that Agelas be transferred from the order Poecilosclerida to the Axinellida.

The methods employed in analysis of sexual reproductive
processes in the Porifera have varied in time and with the
types of species under study. In earlier work on the more
common, shallow-water Ceractinomorpha, detailed histological
investigation of one or only a few specimens during the
period of reproductive maturity was usually sufficient to
demonstrate the various stages in the processes of oogenesis,
spermatogenesis, oocyte feeding and cleavage. This simple
approach is practical only where reproductive processes are
asynchronous and extend over relatively long periods of time
in overlapping phases in individual specimens, such as the
larvae brooding, hermaphroditic Ceractinomorpha. Analyses
of such asynchronous specimens is sufficient to provide
only a highly probable sequence of stages in the processes
by approximation to stages described earlier for the Calcarea, by relating the static stages seen in sections to
similar stages of other organisms where the dynamic processes have been followed directly, and by a minor application of common sense. This approach cannot, however, provide an estimation of the temporal relationships between
stages nor is it likely to resolve critical events of short
duration or sudden occurrence such as fertilization or
meiotic division as noted by Fell ('69).

The time of onset and completion of sexual reproduction, as well as a crude estimation of the time relationships of the processes, have been obtained by histological examination of specimens collected periodically from large, local populations -- usual intervals are monthly, bimonthly or weekly. This approach, if extensive, also provides data from which sexual differentiation may be inferred. Successional hermaphroditism remains difficult to demonstrate rigorously in any asynchronous population. Analysis of the fine time parameters of various stages of the reproductive process are only slightly improved over the earlier method, and events of short duration remain unlikely to be detected. Indeed, if the entire reproductive process or major portions occur synchronously over periods shorter than the sampling interval, as has been suggested for some species (Borojević, '67; Simpson and Gilbert, '73), these processes may be missed altogether.

A more recent approach has been the repetitive sampling of a set of marked specimens held in near-natural conditions in cages suspended in native waters (Diaz, '73). This approach should allow unambiguous documentation of sexual determination of specimens, and where sampling interval is sufficiently short, the time scale of the overall reproductive period. The same failures are to be expected here with respect to fine temporal details. The duration of various stages of the reproductive process cannot be estimated with precision unless the process is synchronous throughout entire specimens. Short duration events are thus expected to remain unresolved using this approach. One can further criticise this approach on the grounds that cage holding and repetitive tissue removal may introduce disruptive influences and obscure the natural course of reproduction. The utility of the method will require documentation of the insignificance of these procedures.

With respect to the asynchronous, brooding, contemporaneously hermaphroditic species, the best we can expect in the extension of our present knowledge of reproductive processes is the increase in the number of observations of sectioned material to complete the missing intervals and events. By enumerating the relative frequency of occurrence of various stages, the relative time spans of these may be derived, but an absolute time scale may only be approximated with difficulty, if at all. With respect to dioecious species, it may be possible to isolate individuals in near-natural conditions in the laboratory and carry out controlled fertilization followed by a very short interval sampling program. Such approaches may prove to be prohibitive in terms of required space, time and equipment. A potentially much more productive alternative is the systematic analysis of species in which sexual processes are synchronized throughout entire individuals or entire local populations. The possibility of detecting major natural events such as release of reproductive products and the potential use of such events as time references from which all other stages of the process can be measured appears to offer obvious advantages when compared to all other available approaches.

Inspection of the range of sexual reproductive characteristics which occur within the phylum (see Fell, '74) allows narrowing the search to sexual patterns most promising for fruitful descriptive investigation and most suitable for ultimate experimental manipulation. These variable characters include: basic pattern (oviparity vs viviparity), sexual determination (dioecious or gonochoristic vs monoecious or hermaphroditic -- contemporaneous, protandric or protogynic), types of products (spermatozoa vs oocytes vs motile larvae), form of release (individual particles vs compound masses) and temporal release pattern (asynchronous vs synchronous -- for individual specimens or entire local populations).

All sponges, as sessile adults, release sperm to effect sexual reproduction. In most brooding Ceractinomorpha, spermatogenesis is synchronous within individual follicles, but not throughout entire specimens (Tuzet and Pavans de Ceccatty, '58; Tuzet et al., '70). Maturation stages are commonly encountered in samples collected over one or several months. In such species, release is not expected to be sufficiently spectacular in either time or density of product to be visually noticable in nature to potential observing divers. Such asynchronous release, furthermore, does not lend itself to experimental analysis or manipulation. Spectacular synchronous release of spermatozoa in nature has previously been reported for both individual specimens and local populations (Reiswig, '70). Species which exhibit such release patterns are more likely to serve as productive experimental subjects.

Viviparous species which brood oocytes and developing embryos also tend to release advanced larvae asynchronously for both individual specimens and populations (Bergquist et al., '70; Fry, '71). Sponges with the most desirable combination of characteristics for detailed descriptive and ultimately experimental work are the oviparous species which release oocytes synchronously throughout entire local populations and which exude the oocytes in a compact gelatinous mass. The gelatinous mass, in which development takes

place, should be of sufficient size to be easily detectable
in nature, should be easily collected in quantity, should be
easily examined microscopically without the complications
usually imposed by parental tissue, should be able to be
fixed for histological examination without the problems of
parental skeletal materials, and finally, such masses should
be easily amenable to experimental manipulation in the field
or laboratory. Such populations are now known and will be
reported below. Two additional characters which an ideal
population would be expected to exhibit are complete sexual
separation and absolute predictability of reproductive
events. The data presented below represent the available
information on gamete release by sponges in nature, and hopefully provide an initial basis for prediction of future release events.

REPORTS OF RELEASE EVENTS

Reports of gamete release originate either from directly
witnessed events or from conditions which obviously indicate
that release occurred a short time prior to the observation
period. The latter consist of discovery of sheets or strands
of gelatinous matrix containing oocytes or developing embryos
strewn over the surfaces of parental sponges of the adjacent
substrate. A total of 25 release events are known to the
author to this date (Table 1). The majority of events are
based on reports of observations generously contributed by
cooperating individuals engaged in sport or research diving
in the Caribbean area. Report data received by the author
typically consists of location of the event, time of day,
date, depth, and a description of the event. In some cases
additional material such as photographic records of release
or preserved samples of the release products are included.
Fortunately the critical bit of information for identification of the parent -- a preserved or dried fragment of the
sponge involved -- has been included with most reports.

SPERMATOZOA RELEASE

Massive release of spermatozoa by one or a few individuals
of a local population has been reported for a variety of
sponges of wide taxonomic distribution (Fig. 1). Such release, always occurring by way of the exhalant canal system
and oscula, obviously represents a synchronous behavioral
event by the releasing specimen. It is not known in any
single case, if such release is preceded by synchrony of
spermatogenesis throughout the specimen. Since the two processes, spermatogenesis and spermatozoa release, involve
quite different controls, synchrony of one does not necessar-

Table 1

Known gamete release events occurring in nature for Caribbean sponges.

Taxonomic position	date dy/mo/yr	time (hours)	locality	gamete (1)	number (2)	depth meters
DICTYOCERATIDA						
Verongia archeri (Higgin)	14/2/69	1500	Jamaica	S	1	49
Verongia cf crassa ? (Hyatt)	28/5/71	1400	Jamaica	S	2	26
DENDROCERATIDA						
Ianthella sp.	--/7/72	--	Jamaica	S	1	--
HAPLOSCLERIDA						
Xestospongia infundibula (Schmidt)	6/4/73	1550	Gd Bahama	S	2	12
Xestospongia muta (Schmidt)	20/8/73	2130	Barbados	S	1	15
POECILOSCLERIDA						
Agelas sp A	29/7/67	1400	Cozumel	S & O	P	31
" " "	27/7/70	--	Gd Cayman	O	P	24
" " "	16/7/71	1700	Jamaica	S	3	28
" " "	24/7/71	1700	Jamaica	O	3	18
" " "	1/8/72	1400	Jamaica	O	-	23
Agelas sp B	23/7/71	1607	Jamaica	S	1	28
" " "	20/4/74	1815	Jamaica	S	P	14
" " "	13/7/74	1620	Jamaica	S & O	4	20
" " "	14/7/74	--	Jamaica	O	P	25
" " "	15/7/74	--	Jamaica	O	1	15
Agelas sp C	15/7/74	1430	Barbados	O	1	12
Neofibularia nolitangere (D&M)	23/10/69	1430	Jamaica	S	P	15
HALICHONDRIDA (none)						
AXINELLIDA						
Hemectyon ferox (D&M)	5/8/69	--	Jamaica	O	1	30
" "	13/8/71	--	Panama	O	P	9
" "	15/8/71	--	Panama	O	P	9
" "	13/9/71	--	Jamaica	O	1	26
" "	30/8/72	1400	Barbados	O	1	14
HADROMERIDA (none)						
SPIROPHORIDA (none)						
CHORISTIDA						
Geodia gibberosa (Lamarck)	2/7/69	1600	Jamaica	S	1	27
HOMOSCLEROPHORIDA (none)						
Unidentified						
brown tubular sponge	30/8/69	--	Curacao	S	1	--
red tubular sponge	16/7/71	--	Panama	O	1	10

(1) abbreviations are S for spermatozoa and O for oocytes.
(2) number of individuals active in release, P represents entire local population.

ily imply synchrony of the other. Synchronization of spermatozoa release throughout local populations has been well
documented for only two species, one of which (<u>Agelas</u> sp A)
is oviparous, while the reproductive pattern of the other
(<u>Neofibularia</u> <u>nolitangere</u>) remains unknown.

Fig. 1. A specimen of
<u>Verongia</u> cf <u>crassa</u> releas-
ing spermatozoa at -26
meters on the reef off
Discovery Bay, Jamaica.
(Photograph contributed
by P. Dustan).

When all spermatozoa reports are considered together, these
appear to have no obvious relationship to season (Fig. 2) or
lunar phase (Fig. 3a). They are, however, obviously clumped
by time of day, between 1400 and 1700 hours local time. The
afternoon concentration is not due to a bias of observer
availability, since habits of Caribbean divers insure at
least equivalent in-water exposure before noon. The period
from midnight to one hour after sunrise can be considered
to be essentially without observation.

OOCYTE RELEASE
Release of female gametes by Caribbean sponges has been reported only for species in which the products are extruded
through the general dermis and enclosed in gelatinous sheets
or strands (Fig. 4). A similar process has been described
earlier for <u>Polymastia</u> <u>robusta</u> by Borojević ('67). The composition of the matrix remains to be investigated. Although
oviparity is well known in many Axinellida and Hadromerida
(Lévi, '56), it is not known whether fertilization takes
place before or after extrusion of the oocytes from parental
tissues. The ability to detect natural release in the oviparous species is attributable to the relatively long period
of time the matrix with developing embryos remains exposed
on the parental surface. Release of oocytes as free particles is unlikely to be noted in nature due to the relatively
short time available for detection and the low attractiveness

of even a synchronous release.

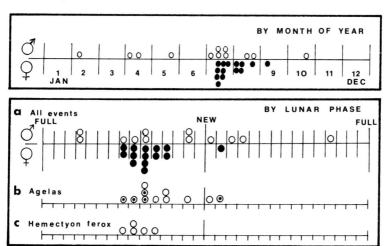

Fig. 2. Seasonal occurrence of all recorded gamete release events of Caribbean sponges (upper graph).
Fig. 3. (below) Relation of gamete release events to lunar phase for a) all events recorded, b) Agelas (sp A with dark centers) and c) Hemectyon ferox.

Fig. 4. A typical post-release condition of Agelas sp A at -24 meters off Grand Cayman. The rear specimens are free of oocyte strands and may be males. The individual in the foreground is a maternal parent enveloped by strands bearing oocytes and developing embryos. Photograph by R.A. Kinzie.

All reports of natural oocyte release in the Caribbean fauna involve only four species of two genera. The oocytes of all of these are nearly identical in size (120 µm in diameter) and color (orange to pink) and are indistinguishable to the unaided eye. The three Agelas forms cannot be satisfactorily assigned to any of the available specific names, including those discussed by Hechtel ('69). Agelas sp A, a hollow, tubular form shown in figure 4, is also illustrated on the cover photograph of the issue containing the Goreau and Hartman ('66) article, and in Hollis ('68). This widely distributed Caribbean form releases both gamete types simultaneously and spectacularly throughout local populations. The sexual condition of the species has not been definitely determined, but all evidence indicates it to be dioecious. The five release events recorded for this form between 1967 and 1972 are closely correlated in seasonality (late July or early August) and lunar phase (Fig. 3b).

Agelas sp B, a dark brown, encrusting plaque, one to several cm in thickness, is known only from Jamaica. Excepting one report of doubtful identification, this species appears to reproduce at the same time, and, as far as is presently known, in a manner identical with that of sp A. Agelas sp C is an orange-tan, repent, branching, cylindrical form with diameter of the solid, slightly irregular branches ranging from 3 to 7 cm. It is clearly distinct from the orange-red A. sceptrum co-occurring with this species in Barbados. The single report indicates that the timing and details of sexual reproduction are again identical to those of the other Agelas forms.

The major stages encountered in Agelas embryogenesis are shown in figure 5. The oocytes have no follicle cell layer, cleavage is complete and regular with ultimate formation of a moderately complex, morphologically polarized, flagellated parenchymula of 1,000 to 2,000 cells. This larva is very similar to that reported from Tethya aurantia by Lévi ('56). Embryogenesis is nearly synchronous in strands from a single specimen. Systematic analysis of the time course of embryogenesis has not yet been attempted.

The other oviparous species, Hemectyon ferox, is widely distributed over the entire tropical western Atlantic region. Only post-release observations (oocyte/embryo strands) have been reported for this species, the time and manner of release have not been directly witnessed for either type of gamete. All post-release observations, however, do indicate a synchronous release which, like that of Agelas, is highly correlated in seasonality and lunar phase from year to year (Fig. 3c). The correlation of release over a range of several thousand miles is startling.

The oocytes of H. ferox differ basically from those of Agelas in the presence of a surrounding layer of cuboidal follicle cells within the gelatinous strings (Fig. 5). Throughout the early stages of regular, equal cleavage, the

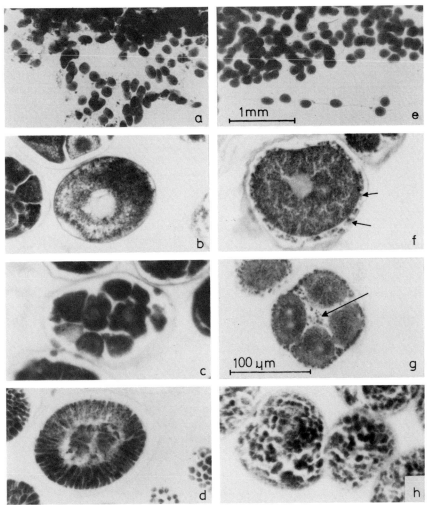

Fig. 5. Developmental stages of <u>Agelas</u> sp A (left) and
<u>Hemectyon</u> <u>ferox</u> (right). Stained whole mounts of oocyte/
embryo strands at equal magnification are shown in a and e.
Sectioned stages at equal magifications are shown for
oocytes (b and f), early cleavage stages (c and g) and the
most advanced stages encountered (d and h). Follicle cells
of <u>H</u>. <u>ferox</u> are clearly seen surrounding the oocyte in f
and invading the cleavage spaces in g (arrows).

follicle cells are found between the blastomeres where they remain easily identifiable. In more advanced stages, as blastomeres approach the follicle cells in size, the two lineages are no longer distinguishable. The ultimate fate of the follicle cells is unknown. The most advanced stage encountered in the few samples available is a solid ball of 100 to 150 cells, entirely lacking in evidence of cellular differentiation and thus lacking the columnar flagellated layer characteristic of sponge parenchymulae. A detailed sequential study of development is needed to determine the final structure of the larva at release and the time course of development.

DISCUSSION

The reports listed here demonstrate the common occurrence of massive spermatozoa release in various demosponge taxa. Although release may be synchronous for individual specimens, most observations indicate that population synchrony is the exception rather than the rule. Species in which individuals release spermatozoa asynchronously are not predictable and thus do not offer sufficiently suitable characteristics to warrant the expense of special concentrated investigation at present. Two populations, Agelas sp A and Hemectyon ferox, however, do exhibit synchronized release as well as a high correlation between release events of successive years. A sufficient number of reports have now been accumulated on the natural gamete release of these to allow inauguration of the next stage of analysis -- prediction of the date and, where data exist, the time of future release. Agelas sp A, and perhaps all three Agelas forms, appear to release between 1500 and 1700 hours on the date of the waning quarter moon occurring in late July or early August. Hemectyon ferox appears to release exactly one lunar month later, at a time of day not yet determined. Data for Neofibularia nolitangere are not yet sufficient to allow prediction of future release events for this species.

Should prediction of future events prove sufficiently accurate, the potential value of these systems for research into basic biological processes of the Porifera is obvious. Specimens may be isolated a day before release to provide separate gamete collections and materials for controlled fertilization. Crosses of known parents would allow a beginning to be made in the investigation of the genetic basis of a variety of characters -- such as gross morphology, spicule dimensions, sex determination, strain specificity, etc. Rapid field collection of released strands of oocytes

would allow description of the full sequence and time course
of embryogenesis, both from direct in vivo observation as
well as from histological analysis of samples fixed at intervals. The range of subjects that would be opened to experimental manipulation is enormous. The following subjects
come to mind among many other possible ones. How is gamete
release controlled within individuals and populations? How
and why is synchrony maintained over distances of thousands
of miles? Are pheromones employed within local populations
to correlate synchrony? How is sperm recognition and transfer to the oocyte effected? What reactions occur between
a sponge and sperm which are received, but which are inap-
propriate in terms of species or strain characteristics?
Why do the three Agelas forms release gametes at the same
time? Is this an ecologically tenable situation or may it
indicate a lack of full species-level separation? What is
the ultimate fate of the follicle cells in Hemectyon ferox?
Are these parental cells incorporated into the larvae render-
ing the next generation a genetic mosaic? Similar problems
of genetic mosaicism have been noted for other larvae
(Simpson, '68; Warburton, '61) and for cases of common larval fusion (Fry, '70, '71). These are only some of the
problems which may be confronted if and when these reproductive systems prove to be reliably predictable.

Another important issue resulting from the reports is the
relationship between oviparity and major taxa of the Demospongiae. Oviparity in Hemectyon ferox is consistent with
its recently accepted inclusion in the Axinellida (Bergquist
and Hartman, '69; Brien et al., '73). Its discovery here
reinforces the general separation of the oviparous
Tetractinomorpha from the viviparous Ceractinomorpha proposed by Lévi ('57). The position of Agelas, usually included within the Poecilosclerida, has always been rather uncertain because of the lack of a close morphological relationship to any other group of extant sponges. Bergquist
and Hartman ('69) have shown that the amino acid pattern of
Agelas is difficult to distinguish from those of axinellids
and is entirely unlike the patterns found in ceractinomorphs.
No sponge which is clearly a valid member of the Ceractinomorpha is known to be oviparous. On the basis of the oviparity of Agelas reported here, it is proposed that the
Agelasidae be removed from the Ceractinomorpha and be included in the order Axinellida. In its new location, the
mono-generic family will still lack close relationship to
any extant member of its order.

Finally, continued documentation of natural reproductive
patterns of the Demospongiae may eventually provide some in-

sight into the ecologically significant selective pressures which operate to maintain the amazing variety in these patterns of the class. At present, detailed information on ecological positions of most Demospongiae is unknown. The few cases where ecological strategies as well as reproductive patterns have been worked out are still insufficient to allow derivation of a hypothesis broad enough to encompass the class as a whole. The factors which maintain viviparous and oviparous species, or asynchronous and synchronous species side by side, in the same habitat, ostensibly utilizing the same resource pool cannot be explained at present. Continued documentation of natural release events are expected to provide us with the best predictive species for immediate future investigation as well as invaluable data on the variety of patterns exhibited by less useful, but, in the long run perhaps, more important species for determining what it is that sponges are doing and why they remain in existence.

ACKNOWLEDGMENTS

I would like to offer my special appreciation to Miss Eileen Graham for her important role in serving as report coordinator and for providing several personal observations on gamete release. I also gratefully acknowledge the cooperative assistance given by Dr. Robert A. Kinzie III, Dr. Lynton S. Land, Dr. Judith Land, Dr. Eugene Shinn, Mr. Carl Roessler, Dr. Phillip Dustan, Dr. David Meyer, Dr. Charles Birkeland, Dr. Peter Spencer-Davies, Dr. Patrick Colin, Mr. Michael Lands, Miss Sharon Ohlhorst, Dr. Elizabeth Sides, Mr. John Gifford, Mr. Peter Reeson, and the many other divers who have contributed to the project over the years.

NOTE ADDED IN PROOF

The predicted 1975 release by <u>Hemectyon ferox</u> has been substantiated by reports of egg strands on 27 - 28 August in Barbados and 31 August 1975 in Jamaica.

LITERATURE CITED

Bergquist, P.R. and W.D. Hartman 1969 Free amino acid patterns and the classification of the Demospongiae. Mar. Biol., 3: 247-268.
Bergquist, P.R., M.E. Sinclair and J.J. Hogg 1970 Adaptation to intertidal existence: reproductive cycles and larval behavior in Demospongiae. Symp. zool. Soc. Lond., 25: 247-271.

Borojević, R. 1967 La Ponte et le développement de Poly-
mastia robusta. Cah. Biol. Mar., 8: 1-6.
Brien, P., C. Lévi, M. Sarà, O. Tuzet and J. Vacelet 1973
Spongiaires. Tome III, Fasc. 1 of "Traité de Zoologie;
Anatomie, Systématique, Biologie" (P.P. Grassé, ed.),
Masson et Cie, Paris, 716 pp.
Diaz, J.P. 1973 Cycle sexuel de deux Démosponges de
l'étang de Thau: Suberites massa Nardo et Hymeniacidon
caruncula Bowerbank. Bull. Soc. Zool. France, 98: 145-
156.
Fell, P.E. 1969 The involvement of nurse cells in oogenesis
and embryonic development in the marine sponge Haliclona
ecbasis. J. Morph., 127: 133-150.
_____1974 Porifera. Chapter 2 of "Reproduction of Marine
Invertebrates, vol. I. Acoelomate and Pseudocoelomate
Metazoans", Academic Press, New York, pp. 51-132.
Fry, W.G. 1970 The sponge as a population: a biometric
approach. Symp. zool. Soc. Lond., 25: 135-162.
_____1971 The biology of larvae of Ophlitaspongia seriata
from two North Wales populations. Europ. Mar. Biol.
Symp., 4: 155-178.
Goreau, T.F. and W.D. Hartman 1966 Sponge effect on the
form of reef corals. Science, 151: 343-344 and issue
cover photog.
Hechtel, G.J. 1969 New species and records of shallow
water Demospongiae from Barbados, West Indies. Postilla,
132: 1-38.
Hollis, B. 1968 Mystery of the smoking sponges. Skin
Diver Mag., 17: 36-39.
Lévi, C. 1956 Étude des Halisarca de Roscoff. Embryologie
et systématique des Démosponges. Arch. Zool. Exp. Gén.,
93: 1-181.
_____1957 Ontogeny and systematics in sponges. Syst. Zool.,
6: 174-183.
Reiswig, H.M. 1970 Porifera: sudden sperm release by
tropical Demospongiae. Science, 170: 538-539.
Simpson, T.L. 1968 The structure and function of sponge
cells. New criteria for the taxonomy of Poecilosclerid
sponges (Demospongiae). Bull. Peabody Mus. Nat. Hist., 25:
1-141.
Simpson, T.L. and J.J. Gilbert 1973 Gemmulation, gemmule
hatching, and sexual reproduction in fresh-water sponges.
1. The life cycle of Spongilla lacustris and Tubella
pennsylvanica. Tr. Amer. Micr. Soc., 92: 422-433.
Tuzet, O. and M. Pavans de Ceccatty 1958 La spermatogénése,
l'ovogenése, la fécondation et les premiers stades du
développement d'Hippospongia communis LMK (=H. equina

O.S.). Bull. Biol. Fr. Belge., 92: 331-348.
Tuzet, O., R. Garrone and M. Pavans de Ceccatty 1970 Observations ultrastructurales sur la spermatogenèse chez la Démosponge Aplysilla rosea Schulze (Dendroceratide): une métaplasie exemplaire. An. Sci. Natur. Zool. Biol. Anim., 12e ser., 12: 27-50.
Warburton, F.E. 1961 Inclusion of parental somatic cells in sponge larvae. Nature, 191: 1317.

REPRODUCTION AND SPECIATION IN HALISARCA

Wen-Tien Chen

Department of Biology
Yale University
New Haven, Connecticut   06520

ABSTRACT: Annual changes in the structure and reproduction of two aspiculiferous sponges, Halisarca dujardini
Johnston and Halisarca nahantensis sp. nov., have been
investigated over a three year period (1972 to 1974) at
Nahant, Massachusetts. The morphology and reproduction
of the new species, Halisarca nahantensis, is described.
These two sibling species are similar in morphology but
show clear ecological and reproductive isolation. Halisarca dujardini inhabits the subtidal zone, while Halisarca nahantensis lives in the intertidal zone. H. dujardini is dioecious and produces mature gametes in
March, while H. nahantensis is monoecious and forms mature gametes in May. The two species are distinct in
their spermatogenesis and in sperm morphology.

Marine slime sponges of the genus Halisarca Dujardin 1838
are widely distributed on the coast of North America. The
specific identity of these sponges has not been published
due to the lack of information on their reproduction and
embryology. Members of this genus lack spicules, and no
morphological structure is constant enough to be applicable
in the determination of species. Halisarca species from
Nahant areas were first introduced to me by Dr. Nathan W.
Riser, Director, Marine Science Institute, Northeastern
University, Boston, in the winter of 1971. I decided to
approach the specific classification of Halisarca by annual
studies of the changes in the structure and reproduction of
sponges collected from eight natural habitats in Nahant.
The results have shown that there are two species, Halisarca
dujardini Johnston and Halisarca nahantensis sp. nov.,
existing along the coast at Nahant. The species show clear
ecological and reproductive isolation but are similar in
histology and anatomy.

MATERIALS AND METHODS

This study was carried out over a three year period, 1972
to 1974 at Nahant, Massachusetts. Samples of sponges were
collected biweekly from June to December and weekly from

113

January to May. Additional samplings were made during embryogenesis of sponges, in April and May. Since the morphology of this sponge is highly plastic, particular attention
was made to sample from the same individual sponges in the
field throughout the year-round studies. Eight large individuals in the natural populations were chosen on the basis
of their appearance (shape, color, and thickness) and habitat. Only a small piece was removed for histological and
embryological studies at each sampling date.

During April and May, most specimens of Halisarca
nahantensis become whitish due to the presence of eggs or
embryos. Some of these sponges that were attached to a
movable substrate, such as a small rock or mussel shell were
removed intact and kept in a running sea water tank in the
laboratory. The embryos in these sponges showed the same
development as those of field specimens during a one month
culture period.

For histological investigation, specimens were fixed immediately after collection with Zenker's fluid for eight
hours, dehydrated through an ethyl alcohol series, embedded
in polyester wax (Steedman, '60), sectioned serially at 2,
4, 8, or 24 µm, and stained with Heidenhain's hematoxylin
and eosin, or Heidenhain's azan (Gabe, '68).

RESULTS

Halisarca nahantensis sp. nov.

Holotype: Peabody Museum of Natural History, Yale
University (YPM) No. 9034. Specimen collected 0.8
meter above mean low water at Nahant, Massachusetts,
U.S.A., on April 24, 1974.
Repositories of other type material: Marine Science
Institute, Northeastern University; YPM Nos. 9035 to
9044; and National Museum of Natural History,
Washington, D.C.

Diagnosis: Inhabiting marine intertidal zone; spherical
cells with equal cytoplasmic inclusions, sometimes with
additional small inclusions; haploid chromosome number 22;
monoecious; spermatogenesis asynchronous; sperm with a
lemon-shaped head; larva white and large, about 150 µm in
diameter, completely flagellated, with a folded invaginated
cavity; modified rhagon.

Habitat: H. nahantensis occurs intertidally in the rocky
shore area. Specimens encrust rocks, shells of Mytilus

edulis and Modiolus modiolus, and the alga, Chondrus cris-
pus. H. nahantensis grows on the sides and undersides of
rocks where it is hidden from direct exposure to sunlight.
It is often associated with the calcareous sponge Leucosolenia sp.

Morphology: H. nahantensis forms thin encrustations on the
substrate. Its size ranges up to 15 cm in diameter and 1 to
4 mm in thickness. Tubular oscula are sometimes present,
varying in height up to 1 mm with basal diameters of 0.5 to
1 mm. Color in life is whitish to brownish. Often the same
individuals will show gradation from a lighter to a darker
color. The thinner individuals containing ova or embryos
are whitish, but the thicker individuals are often tan. The
consistency is soft and gelatinous, while the surface is
smooth and mucoid.

The anatomy, histology, and cytology of H. nahantensis is
basically the same as that of other members in this genus as
recorded by Lévi ('56). A discussion of the annual changes
in structure correlated with the reproductive cycle is given
below. The cell types, which have been previously described
by Topsent (1893), Tuzet and Pavans de Ceccatty ('55), and
Lévi ('56) occur within the mesohyl of H. nahantensis.
Spherical cells, "cellules sphéruleuses" of Lévi ('56), tend
to be localized near the ectosome or the aquiferous canals.
Topsent (1893) and Lévi ('56) regarded the morphology of
spherical cells as one of the specific characters of H.
sputum and H. metschnikovi respectively. The spherical cells
of H. nahantensis measure 10 to 12 μm in diameter, with a
nucleus of 2 to 3 μm, and contain 8 to 14 cytoplasmic inclusions. The size of the inclusions varies from 2 to 5 μm
in diameter. These inclusions are circular and are grouped
like clusters of grapes.

Reproduction: During 1972 and 1973, the reproductive season
of H. nahantensis continued from November through June (Fig.
1). Spermatogenesis occurs from late March through May and
oogenesis occurs from November through May. The larvae
appear in May and June. There is only one cycle of gametogenesis per year. The oocytes appear four months before the
spermatocytes, and only after the oocytes reach the late
stage of vitellogenesis does spermatogenesis begin. During
May, mature gametes are found in decreasing numbers.

H. nahantensis is monoecious. Specimens collected in
April contain both spermatic cysts and oocytic cysts. During early May some specimens contain only mature ova, probably due to the prior release of sperm. The spermatic cysts

and the oocytic cysts of H. nahantensis are intermingled in the basal two-thirds of the sponge. Oogenesis and development of embryos in H. nahantensis are generally synchronous within individuals. Although most of the specimens examined at a particular time and locality contained oocytes or embryos at the same stage of development, the marginal portion of some specimens contained oocytes or embryos younger than those of the central portion. However, spermatogenesis in H. nahantensis is asynchronous. Different developmental stages of the male cell line are found in the same spermatic cyst, as well as in different regions of the same sponge individual.

Embryology: Living spermatozoa of H. nahantensis consist of a lemon-shaped head, 5 μm x 4 μm x 4 μm, and a long tail, 30 to 40 μm in length (Fig. 2H,I). In histological preparations (Fig. 2J,K), the spermatozoan is characterized by a cone-shaped nucleus with chromatin condensed in the pointed end.

The young oocytes of H. nahantensis measure about 50 μm in diameter during the initial stage of vitellogenesis. When the cells reach about 100 μm in diameter, the cytoplasm of the oocyte is fully filled with yolk granules and the filiform cytoplasmic processes become the pseudopodia. The mature oocyte is enclosed in an endopinacocyte-lined oocyst. The oocytic cysts are circular, sometimes oval, and measure 120 to 130 μm in diameter. Before undergoing maturation the oocytes contain a large nucleus, ranging up to 40 μm in diameter with a nucleolus 10 μm in diameter.

In somatic cells of Halisarca, the chromosomes are too small to be characterized in histological preparations. The oocyte during maturation divisions contains a large nucleus and chromosomes that measure from 0.2 to 0.4 μm in length and can be counted in serial sections (4 μm thick polyester sections). During metaphase I, the equatorial plates have an average diameter of about 8 μm and in sections vertically cut at 4 μm, most of the chromosome pairs appeared in three sections. Sometimes a cross section of the oocyte contained the whole equatorial plate. The number of chromosome pairs in the oocyte of H. nahantensis (from 20 samples) is 22. This indicates that H. nahantensis has a diploid chromosome number of 44.

H. nahantensis is viviparous. The zygote goes through its early development, up to the formation of the larva, within the oocytic follicle. Cleavage starts from a normal radial cleavage at the 2-cell stage. After the 16-cell stage, the embryo consists of equal blastomeres arranged radially with a small blastocoel.

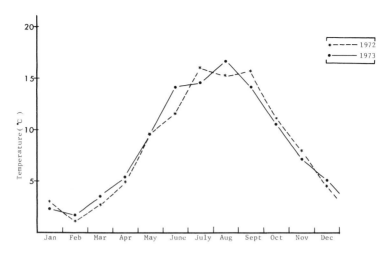

Fig. 1.A. Monthly means of the intertidal surface sea water
temperature in Nahant, Massachusetts, from 1972 to 1973.
Temperature readings are taken from records kept in the
Marine Science Institute, Northeastern University, Nahant,
Massachusetts, U.S.A.

The larva of H. nahantensis forms an internal pocket pro-
jecting into the interior through lateral invagination during its late nurse stage. At the same time, the anterior
cells proliferate actively. The invaginated pocket is usually folded and measures 30 to 50 µm across the widest portion. According to Lévi ('56, '63), the internal cavities
found in the larvae of H. dujardini and H. metschnikovi are
formed by the uni- and multipolar migration of external
flagellated cells that associate and become an internal cellular mass, instead of by invagination of lateral cells as
in this study.

The hatching larvae of H. nahantensis are white and oval-
shaped. The free-living larvae are modified from oval to
flat bell-shaped. The oval larvae measure from 150 to 200
µm in length and 120 to 160 µm across the widest portion,
while the flat forms are 120 to 140 µm in length and 140 to
180 µm in width. They are the largest Halisarca larvae yet
described: the larvae of H. dujardini measure 90 to 95 µm in
length, while those of H. metschnikovi are 120 to 130 µm.

The larva of H. nahantensis fixes to the substratum by the
invaginated side, and undergoes a complete change of form
and appearance, becoming a white flattened mass which contains a compact mass of histologically identical cells. The

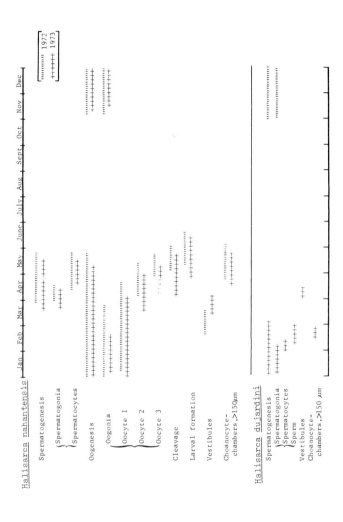

Fig. 1. B. Annual reproductive cycles in H. nahantensis and H. dujardini. Only data from sponge individuals undergoing sexual reproduction during the two-year period are recorded in this figure. Oocyte 1, initial growth stage of the oocyte; Oocyte 2, vitellogenetic stage of the oocyte; Oocyte 3, follicular stage of the oocyte.

peripheral cells of this compact cell mass are taper-shaped.
This stage is considered as a "diamorph" stage. By the end
of the third day of larval attachment at 13°C, a young sponge
develops a more or less tear drop shape with a tubular osculum on the top. This young sponge of H. nahantensis is considered as a modified rhagon.

H. nahantensis Compared With H. dujardini

Ecology and Reproduction: At Nahant, Massachusetts, the
genus Halisarca includes the species, H. dujardini Johnston
1842 as well as H. nahantensis. These forms occur from about
one meter above mean low water to depths of at least 4 meters
in the rocky shore area. The majority of intertidal individuals are H. nahantensis, often associated with the calcareous sponge, Leucosolenia sp. The subtidal species, H. dujardini, often coexists with Clathrina sp. Both species of
Halisarca grow on the sides and the undersides of rocks where
they have no direct exposure to sunlight. It is nearly impossible to distinguish the two species by their external
morphology. However, it may be noted that H. dujardini
growing on algae tends to be globular in form, while H.
nahantensis is flat.

H. nahantensis and H. dujardini show clear reproductive
isolation in the natural habitat. The gametes of H. nahantensis mature in May, while those of H. dujardini mature in
February. The monthly means of intertidal sea water temperature in Nahant and the annual reproductive cycles in both
species of Halisarca during the two-year study period are
presented in figure 1. Both species breed annually, with
various reproductive events being restricted to specific periods. Following sexual reproduction, there is a period, summer and fall, during which both sponges undergo active
growth and contain no sex products. No gemmule formation was
observed in these sponges during the two-year period of
study.

H. dujardini in Nahant is dioecious. Among 24 specimens
examined from November, 1972 to April, 1974, those that were
sexually mature either contained only male gametes or only
female gametes. Spermatogenesis is not only synchronous
within individual specimens but also among the individuals of
the population. The spermatogonial stages occur from November to December. The spermatic cysts are found in January
and February. Sperm are released in late February. Unfortunately, complete data on oogenesis and embryonic development of H. dujardini in Nahant is lacking because of the
severe difficulties in sampling under icy winter water.

Date	11/18/72	1/10/73	2/15/73	3/31/73	4/26/72	4/13/73	4/16/72
Water temperature (°C)	6.8	0.1	0.8	3.8	5.4	2.9	5.8
Reproductive stages	oogonial	oocyte-1	oocyte-1	oocyte-2	oocyte-2	oocyte-2	oocyte-3
Ectosome	+	+	+	+	+	+	+
Aquiferous canals	++	++	++	++	++	+	-
Oscula	++	++	+	+	++	+	-
Ostia	-	++	+	+	++	+	-
Vestibules	-	++	+	+	++	+	-
Choanocyte-chambers	+++	+++	+++	+++	+++	++	-
Mesohylial vasculization	-	-	-	+++	-	+++	-

Date	5/5/72	5/18/72	5/31/73	6/16/72	8/9/72	10/14/72
Water temperature (°C)	7.7	8.8	12.4	10.8	14.7	11.1
Reproductive stages	early cleavage	late cleavage	larvae	growth	growth	growth
Ectosome	+	+	+	+	+	+
Aquiferous canals	+	+*	+++	+++	+++	++
Oscula	-	-	+++	+++	+++	++
Ostia	-	-	++	+++	+++	++
Vestibules	-	-	++	+++	+++	++
Choanocyte-chambers	+	++	++	+++**	+++**	+++
Mesohylial vasculization	-	-	-	-	-	-

Table 1. Summary of annual collection of adult Halisarca
nahantensis during 1972–1973 showing structural changes with
respect to temperature changes and developmental stages.
* A few aquiferous canals are observed, but most of them
   are fully filled with migrating amoeboid cells.
**Most choanocyte-chambers are longer than 400 μm.   The
   length of choanocyte-chambers is up to 600 μm.   The width
   of choanocyte-chambers ranges from 25 μm to 100 μm.

However, since the timing of spermatogenesis in H. dujardini
of Nahant is known, it can be predicted that they will produce mature ova in February or March, and release larvae in
April. This inference is based on the following facts:
(1) sperm production occurs during late February, (2) in the
intertidal species, H. nahantensis, sperm and egg production
take place at the same time, and larval production occurs
one month later, and (3) some fragmentary information on the
oocytes and the embryos of the subtidal Halisarca has been
obtained.

Morphology of Sexual Halisarca: The structure and cell types
of both Halisarca species change throughout the year with
respect to the reproductive cycle. Cellular changes are

closely associated with structural changes in both species:
the reduction of choanocyte chambers is accompanied by an increase in the number of "nucleolate" cells. The growth phase
of both species reveals functional structures: aquiferous
canals, oscula, ostia, vestibules, and choanocyte chambers;
while during the reproductive phase of both species, the
functional structures are somehow reduced.

The structural and cellular changes in H. nahantensis
throughout the year with respect to the reproductive cycle
and temperature are represented in Table 1. H. nahantensis
specimens collected from June through February are fully
functional sponges. A complete aquiferous system and choanocyte chambers are found in the sponge. The ostia are usually
associated with the vestibules. Since the ostia and the
vestibules only appear occasionally, the presence of both
structures might depend on physiological conditions of the
sponge. The ostia measure from 5 to 20 μm in diameter in
histological preparations. The vestibules measure up to
150 μm in diameter. During March and April, when the oocytes
undergo vitellogenesis and the spermatogonia appear, aquiferous canals and choanocyte chambers are generally reduced
(Fig. 4). By the time mature ova and spermatic cysts are
forming, only traces of choanocyte chambers remain (Fig. 6).
The aquiferous system, comprising oscula, ostia, vestibules,
and aquiferous canals, is not present during this stage.
Whether the choanocyte chambers will completely disappear has
not been ascertained. However, in one out of twentyfour
specimens examined, all remnants of the aquiferous system and
the choanocyte chambers were absent leaving only the various
amoeboid cells, mature ova, and spermatic cysts scattered
loosely throughout the sponge. During May, when embryos and
larvae are forming, aquiferous canals and choanocyte chambers
progressively appear in the mesohyl of adult sponges (Fig.7).
A fully functional aquiferous system is found during larval
release, while fully functional choanocyte chambers appear
only after larval release.

In H. dujardini, the reduction of aquiferous canals and
choanocyte chambers occurs slowly and regularly with respect
to the long period of synchronous spermatogenesis. Figures
8 to 11 show the general histology of small pieces from the
same individual collected on 11/18/1972, 1/10/1973, 2/15/1973,
and 4/15/1973 respectively. During November, the aquiferous
canals and the choanocyte chambers were progressively reduced. Some choanocyte chambers appear to be closed (that
is, showing no evidence of connection with the aquiferous
system) in histological preparations. During January the
aquiferous system including the ostia and oscula disappeared

and the choanocyte chambers were reduced and closed. During
that time many "nucleolate" cells and spermatogonia occupy
the basal two thirds of the mesohyl. Cells in some previous
choanocyte chambers are indistinguishable from groups of "nucleolate" cells. In many previous choanocyte chambers trans-
formation from choanocytes to "nucleolate" cells can be observed. The choanocytes first appear to lose their collars,
round up, and lose their flagella. It is not certain whether
the "nucleolate" cells from choanocytes give rise to the
gonia or whether preexisting "nucleolate" cells in the mesohyl become the gonia.

H. dujardini collected in February contains many spermatic
cysts in the basal two thirds of the sponge. The aquiferous
system appears and the choanocyte chambers are open and
functioning (Fig. 10). In some samples of H. dujardini collected in March and April, the mesohyl appears to be composed
of numerous contiguous channels separated by delicate membranous partitions. I shall refer to this condition of the
mesohyl as vascularization. The cause of this phenomenon is
unknown. The release of sperm does not seem to be the cause,
since some samples of H. nahantensis collected in March and
April contained oocytes and sperm, and also showed the vascularization (Fig. 5).

Among structures recognized, only the ectosome is constant
throughout the year in either H. nahantensis or H. dujardini.
The ectosome of both H. nahantensis and H. dujardini consists
of three layers. The outer layer (the gelatinous zone) varies greatly in thickness up to 10 μm and contains amoeboid
cells and foreign particles embedded in a gelatinous matrix.
The middle layer (the fibrous zone) measures 5 μm in thickness and contains mostly fibrous structures which stain with
alcian blue, PAS, aniline blue, and toluidine blue. Herlant-
Meewis ('48) suggested that fibers found in H. dujardini are
collagenous on the basis of aniline blue and toluidine blue
staining, but a more certain identification can only be made
by electron microscopy. The inner layer (the cellular zone)
is composed of exopinacoderm and gland cells, occupying a
region of 10 to 20 μm in thickness. Long cytoplasmic processes of gland cells extend outward through the exopinacoderm,
fibrous, and gelatinous zones.

In both species of Halisarca, the size of choanocyte chambers varies greatly from growth to reproductive phases of the
sponges. During the growth phase the choanocyte chambers,
from 250 to 600 μm in length, of both species are grouped in
the form of branched tubules around the exhalant canals.
Choanocyte chambers of 300 to 400 μm in length are commonly
found. The width of choanocyte chambers ranges from 24 to

100 μm. During the reproductive phase the choanocyte chambers of both species are closed and are reduced to about 100 μm in length as described above. The morphology of individual choanocytes changes from time to time during the year. The choanocyte in functioning choanocyte chambers is about 8 μm long and 4 μm wide with an apical nucleus, a flagellum and a collar. The nucleus, about 3 μm in diameter, contains a round nucleolus, about 1 μm in diameter. In a closed choanocyte chamber the collars of choanocytes are not observable. The choanocytes and "nucleolate" cells of H. nahantensis collected during larval release are rounded up and contain numerous cytoplasmic inclusions (Fig. 7).

Two other cell types, "nucleolate" cells and fuchsinophil cells, are closely related with the reproductive cycle of Halisarca in Nahant. "Nucleolate" cells, amebocytes I and II of Lévi ('56), are found throughout the mesohyl. They are regarded as a morphologically homogeneous group of cells in this study. The "nucleolate" cells of both species of Halisarca have various shapes; they can be fusiform or amoeboid. Fuchsinophil cells are scattered throughout the mesohyl and vary greatly in shape. Cell, nucleus, and nucleolus measured 10, 3, and 1 μm in diameter, respectively, and are similar to those of "nucleolate" cells. The fuchsinophil cells contain many cytoplasmic granules, approximately 1 μm in diameter, which become strongly basophilic after being fixed with Zenker's or San Felice's fluid. Dense aggregates of fuchsinophil cells are often found around foreign objects in the sponge, such as pieces of dead alga, nematodes, or sediment. Some cells, which contain only a few granules, but are intermediate between fuchsinophil cells and "nucleolate" cells were observed. It is possible that fuchsinophil cells are specialized "nucleolate" cells which have defensive or transport function.

Tables 2 and 3 show the approximate number of fuchsinophil cells and "nucleolate" cells respectively found in an area, 100 x 100 x 8 μm^3, around three representative structures, ectosome, choanocyte chamber, and oocyte or embryo, during annual collections of adult H. nahantensis. Estimates were made by counting the number of nuclei of either fuchsinophil cells or "nucleolate" cells in 100 μm^2 using an 8 μm thick section. Since the nuclei of fuchsinophil cells and "nucleolate" cells have an average diameter of about 3 μm and the sections were cut at 8 μm, most of the nuclei should appear in one section. Each of these cell types has only one nucleus, ranging from 8 to 14 μm in diameter. Each number in the table represents the average of four measurements. Estimates of the cell number around oocyte or embryo were made by

Date	Reproductive stages	Ectosome	Choanocyte-chamber	Oocyte or embryo
11/18/72	oogonia	3	6	4
1/10/73	oocyte-1	2	5	4
2/15/73	oocyte-1	4	5	2
3/31/73	oocyte-2	2	6	2
4/26/72	oocyte-2	2	4	3
4/13/73	oocyte-2	2	3	2
4/16/73	oocyte-3	3	3	6
5/5/72	early cleavage	1	2	5
5/18/72	late cleavage	3	8	22
5/31/73	larva	1	2	12
6/19/72	growth	1	3	–
8/9/72	growth	2	3	–
10/14/72	growth	3	6	–

Table 2. Summary of annual collection of adult *Halisarca nahantensis* during 1972-1973 showing the approximate number of fuchsinophil cells found in an area, 100 x 100 x 8 μm^3, around three representative structures, ectosome, choanocyte chamber, and oocyte or embryo.

Date	Reproductive stages	Ectosome	Choanocyte-chamber	Oocyte or embryo
11/18/72	oogonia	29	41	39
1/10/73	oocyte-1	30	32	35
2/15/73	oocyte-1	26	31	28
3/31/73	oocyte-2	16	33	119
4/26/72	oocyte-2	19	31	91
4/13/73	oocyte-2	22	29	87
4/16/73	oocyte-3	28	21	33
5/5/72	early cleavage	29	27	26
5/18/72	late cleavage	17	24	25
5/31/73	larva	12	16	22
6/19/72	growth	25	33	–
8/9/72	growth	39	43	–
10/14/72	growth	42	47	–

Table 3. Summary of annual collection of adult *Halisarca nahantensis* during 1972-1973 showing the approximate number of "nucleolate" cells found in an area, 100 x 100 x 8 μm^3, around three representative structures, ectosome, choanocyte chamber, and oocyte or embryo.

counting the number of nuclei in four regions, 100 x 25 μm^2, surrounding the oocyte or embryo. The total area of these four regions is 100 x 100 μm^2 which is comparable to that of other estimates.

Specimens of H. nahantensis with embryos or larvae contain more fuchsinophil cells (Table 2), most of which aggregate around the embryos and migrate into them (see Chen, '74, '76 for morphological evidence). The fuchsinophil cells are therefore maternal nurse cells that become involved in the development of the larvae. These observations are at variance with the work of Lévi ('56) who believed that fuchsinophil cells develop from blastomeres very early in larval formation. The "nucleolate" cells are most abundant in the mesohyl when the oocytes undergo vitellogenesis and the choanocyte chambers start to break up (Table 3). At that time there are dense aggregates of "nucleolate" cells surrounding the oocytes. The oocytes extend their lobe-like cytoplasmic processes radially to contact the filiform processes of the "nucleolate" cells, suggesting that the "nucleolate" cells are involved in vitellogenesis.

Spermatogenesis: In H. nahantensis, spermatogenesis occurs for a short period of approximately 4 weeks (Fig. 1), and is asynchronous. The head of the spermatozoan is lemon-shaped. On the other hand, H. dujardini has a synchronous spermatogenesis which occurs over a longer time period of about 4 months (Fig. 1) and produces sperm with a disc-shaped head.

During the early stage of spermatogenesis in both species, the spermatogonia appear to be isolated in the mesohyl. They are closely associated with "nucleolate" cells in later stages. Spermatogonia are similar to "nucleolate" cells but larger in size, about 10 μm in diameter, and are dividing. The nuclei of spermatogonia measure 4 to 5 μm in diameter and have a nucleolus about 1 to 2 μm in diameter. The nucleus of "nucleolate" cells is 3 μm in diameter; the nucleolus, 1 μm in diameter. There are apparently several spermatogonial generations in H. dujardini where spermatogonial activity extends over a long period of about 3 months (Fig. 1). During late stages, the mesohyl of male H. dujardini is populated with large numbers of spermatogonia (Fig. 9). In H. nahantensis, the isolated spermatogonia divide and aggregate in the mesohyl to form the rudiments of the spermatic cysts (Chen, '74). The number of spermatogonia in H. nahantensis, when compared with that in H. dujardini,is relatively small. This is probably due to the coexistence of male and female gametes in the same individual of H. nahantensis.

Following the spermatogonial generations, the spermatocytes

are surrounded by a continuous layer of follicle cells in both species. The "nucleolate" cells, which are often found among the gonial aggregates during late spermatogonial stages, might give rise to the follicle cells, but this is not certain. In H. dujardini, spermatogenesis in any one cyst is synchronous. During the early stage, the cells measure 8 μm in diameter. As the cells develop, their size is reduced to 5 μm. These two groups of cells are here considered as the primary spermatocytes and the secondary spermatocytes respectively. At a later stage, differentiation of the flagellum is observed in the spermatids which are small cells, 4 μm in diameter. The fine details involved in nuclear division in spermatogenesis are obscure because of the small size of the chromosomes and long developmental period. However, it seems reasonable to assume that spermatogenesis in H. dujardini follows the usual course of two meiotic divisions.

In H. nahantensis, spermatogenesis in any one cyst is asynchronous. After the initial formation of the cyst, all the developmental stages, from spermatocytes to sperm, are usually found in the same cyst. Three groups of cells in the spermatic cyst having various sizes, 10, 7, and 5 μm in diameter, could be referred to as primary spermatocytes, secondary spermatocytes, and spermatids respectively.

In both species of Halisarca, complete condensation of nuclear material and elimination of cytoplasm are not observed in the spermatozoa. The aquiferous canals of both species contain sperm which are actively swimming and contain partially condensed nuclei (Fig. 2 A-K). Whether the spermatozoa complete nuclear condensation and cytoplasmic elimination some time after release is unknown.

H. dujardini release sperm during February and March. The spermatozoa (Figs. 2 A-G, 3) move with a quick, slight jerk and with the point of attachment of the flagellum forward. The head of living spermatozoa is disc-like and measures 5 x 4 x 1.5 μm^3. The tail ranges from 40 to 50 μm in length. In smear preparations (Fig. 2 E-G), the spermatozoan is characterized by a cup-like structure which is stained with hematoxylin but not with Feulgen's nuclear stain. In living sperm (Figs. 2 A-D, 3), a mass of granules occupies this cup-like region. These granules were not demonstrated by the vital stains, Janus green B or neutral red. Whether the cup-like structure is an acrosome or a mitochondrion is unknown. Ultrastructural study will be very helpful in clarifying this situation.

DISCUSSION

It has been known that very few morphological characteristics of sponges, other than those of the skeleton, are
constant enough to be applicable in classifying sponges.
Attempts to identify species of the genus Halisarca by
morphological features have led to a confusing multiplicity
of species. Halisarca is characterized by large branched
choanocyte chambers and a smooth surface. Spicules are absent. Due to the lack of obvious skeletal and morphological
characteristics, the validity of various species in this
genus has been questioned by several authors, who regarded
all Halisarca as conspecific with Halisarca dujardini as
originally described by Johnston (1842).

Many attempts have been made to separate a variety of
species from the typical species, H. dujardini. The descriptions provided by early authors often fail to enable
one to identify the species. In some instances, it is even
uncertain if the organisms described are sponges. Some
species were described on the basis of color. The color of
H. dujardini is whitish to yellowish-brown, but that of H.
magellanica Topsent 1901 and H. purpura Little 1963 is recorded as purplish-red. It may be noted that, if Halisarca
nahantensis is cultured in the laboratory, it becomes green
in water containing the unicellular alga, Dunaliella salina,
orange in that with Monochrysis lutheri, and grey in that
containing Isochrysis galbana.

Anatomical, histological, and cytological data have also
been applied to distinguish several species from the typical
species, H. dujardini. De Laubenfels ('32, '48) established
H. sacra from the coast of California on the basis of its
extraordinarily long and large choanocyte chambers, measuring 200 to 280 μm in length. He stated that the choanocyte
chambers of H. dujardini are commonly 25 μm in diameter by
60 to 150 μm in length. Annual studies of morphological
changes might be useful in judging this criterion since the
anatomy and histology of sponges of this genus are not constant throughout the year. Choanocyte chambers of both
species of Halisarca, collected in various seasons in Nahant, ranged from 50 to 600 μm in length. Sexually mature
specimens are filled with oocytes and sperm, and choanocyte
chambers nearly disappear.

Topsent (1893) established H. sputum on the basis of the
presence of spherical cells. Both the position of the
nucleus in the cell and the number, size, and shape of the
cytoplasmic inclusions are important in his identification.
However, various histological preparations utilizing dif-

127

ferent fixatives and stains might bring out different pictures. Moreover, the morphology of spherical cells is not constant even in the same individual, nor is their function known. It is possible that spherical cells are involved in excretion as suggested by Lévi ('56). On the other hand, they might be symbiotic unicellular organisms, since they are not found in young sponges in the laboratory (Chen, '74, '76).

Anatomical, histological and cytological characters of Halisarca studied at Nahant are basically the same as those found by Lévi ('56), suggesting that such characters are useful in defining the genus.

On the basis of developmental studies, Lévi ('53, '56) distinguished an estuarine form, H. metschnikovi from the typical marine littoral species, H. dujardini. It is rather difficult to apply embryology in distinguishing the sibling species of Halisarca due to the long period of development. Precise information on reproduction and the morphology of certain final stages of development could possibly be useful in species comparisons. The former would include the mode of spermatogenesis and oogenesis, reproductive seasonality and chromosome number. The latter would include the morphology of the spermatozoan, ovum, blastula, free-living larva, and rhagon. Since the shape of the larva is always slightly modified during its free-living stage, larval morphology of both the early free-living stage and the late stage should be recorded.

The pattern of gametogenesis does not seem to be related to the systematic position of sponges. Asynchronous spermatogenesis has been reported by Lévi('56) in Oscarella lobularis and H. metschnikovi, by Tuzet and Paris ('64) in Octavella galangaui, and in this study in H. nahantensis. Synchronous spermatogenesis has been found by Lévi ('56) and confirmed in the present study in H. dujardini, by Tuzet and Pavans de Ceccatty ('58) in Hippospongia communis and by Tuzet et al. ('70) in Aplysilla rosea. In H. dujardini and H. metschnikovi (Lévi, '56), and H. nahantensis oogenesis is synchronous. In Oscarella lobularis the oocytes develop a-synchronously according to Lévi ('51).

Major processes of oogenesis in the genus Halisarca are practically identical. The oocytes of H. dujardini are smaller than those of other members of the genus (Lévi, '56). In H. dujardini, oocytes with two or more nuclei are present in the beginning of the initial growth phase, while in H. nahantensis abnormal oocytes are found in the vitellogenic phase. The ratio of cytoplasm to nucleus in the oocytes of H. metschnikovi is smaller than that of H. nahantensis.

128

This should be accepted with caution, since measurements of the oocytes in living conditions and in histological prepara- tions are different. I do not know which measurement Lévi's data are based on.

Table 4. Biological differences among members
of genus Halisarca

Characteristic compared	H. dujardini Johnston 1842	H. metschnikovi Lévi 1953	H. nahantensis Chen 1974
Habitat	Marine subtidal	Estuary	Marine inter- tidal
"Cellules spheruleuses"	Equal cytoplas- mic inclusions	With one large cytoplasmic inclusion	Equal cytoplas- mic inclusions
Sexuality	Dioecious(?)	Monoecious	Monoecious
Spermato- genesis	Synchronous; long period	Asynchronous; short period	Asynchronous; short period
Sperm	Disc-shaped head	?	Lemon-shaped head
Haploid chromo- some number	?	30	22
Egg size (μm in diameter)	90	120	120 to 130
Larva	Completely flagellated	Posterior pole not flagellated	Completely flagellated
Larval size (μm in diameter)	90	130	150
Larval cavity	Simple	Folded	Folded
Rhagon	Simple	Choanocyte- chamber branching	Choanocyte- chamber branching

Gametogenesis and sperm morphology in H. nahantensis are similar to that seen in H. metschnikova (Lévi, '53, '56). However, in H. nahantensis the number of chromosome pairs es- timated from the oocyte during metaphase I is 22, while the number in H. metschnikovi is about 30. Presumably a complete haploid chromosome set in Halisarca contains 11 chromosomes. H. nahantensis might be a tetraploid, having 44 chromosomes, while H. metschnikovi might be a hexaploid, having 66 chromo- somes.

Among the species of Halisarca previously described along the Atlantic and the Pacific coasts of America, no informa- tion on reproduction nor embryology in H. sacra de Laubenfels has been published. The morphology of H. purpura Little 1963, in the Florida Gulf coast, is distinguished from other members of Halisarca by a thick (600 to 700 μm) alveolar zone in the ectosome and short, sac-like choanocyte chambers. It is suggested that this species could belong to a genus other than Halisarca.

Ecological and reproductive data are useful to separate the sibling species of Halisarca in the same geographical locality. On the American coast at Nahant, ecological distributions of H. nahantensis and H. dujardini overlap. H. nahantensis is dominant in the intertidal zone, while H. dujardini is a subtidal species. Both species are identical in morphology but show clear reproductive isolation. H. nahantensis releases sperm in April, while H. dujardini produces sperm in February. Embryological studies have also demonstrated that the two species are distinct in spermatogenesis and sperm morphology.

Among European species of Halisarca, H. dujardini Johnston 1842 and H. metschnikovi Lévi 1953 were both well described embryologically in Lévi's ('56) work. Table 4 summarizes biological differences among three members of the genus Halisarca. H. dujardini is separated from H. metschnikovi and H. nahantensis by its dioecious nature and by the pattern of spermatogenesis. H. nahantensis differs from H. metschnikovi in chromosome number, larval morphology, sperm and spherical cell structure and natural habitat.

ACKNOWLEDGMENTS

This study represents part of a thesis submitted to the faculty of Northeastern University in partial fulfillment of the requirements for the degree of Master of Science in Biology. I wish to express my appreciation to Dr. Nathan W. Riser, Director, Marine Science Institute, Northeastern University, for his encouragement and direction of this study and to Dr. Patricia M. Morse and Dr. Charles H. Ellis, Jr., for their numerous suggestions and considerable labors on my behalf. Special thanks go to Dr. Willard D. Hartman, Department of Biology and Peabody Museum of Natural History, Yale University, for his critical review and discussion of the manuscript. Contribution No. 40 Marine Science Institute, Northeastern University.

LITERATURE CITED

Chen, W. 1974 The reproductive biology, development, and speciation in the genus Halisarca Dujardin 1838 (Porifera) in Nahant. Thesis, Master of Science Degree, Northeastern University.

_____1976 Oogenesis, embryonic development, and metamorphosis in the marine slime sponge, _Halisarca_ _nahantensis_. J. Morph., (in preparation).

Dujardin, F. 1838 Observations sur les éponges et en particulier sur la Spongille ou éponge d'eau douce. Ann. Sci. Nat., t. S, 2e Serie: 5-13.

Gabe, M. 1968 Techniques histologiques. Masson et Cie, Editeurs, Paris.

Herlant-Meewis, H. 1948 Contribution à l'étude histologique des spongiaires. Ann. Soc. Roy. Zool. Belg., 79: 5-36.

Johnston, G. 1842 A history of British sponges and lithophytes. Edinburgh, W. H. Lizars.

Laubenfels, M.W. de 1932 The marine and freshwater sponges of California. Proc. U.S. Nat. Mus., 81: 23-24.

_____1948 The order Keratosa of the phylum Porifera---A monographic study. Allan Hancock Found. Publ. Occ. Pap. No. 3.

Lévi, C. 1951 Existence d'un stade grégaire transitoire au cours de l'ovogénèse des spongiaires _Halisarca_ dujardini (Johnst.) et _Oscarella_ _lobularis_ (O.S.). C.R. Acad. Sci., Paris, 233: 826-828.

_____1953 _Halisarca_ metschnikovi, n. sp., éponge sans squelette des côtes de France: ses caractères embryologique. Arch. Zool Exp. Gén., 90: 87-91.

_____1956 Étude des _Halisarca_ de Roscoff. Embryologie et systématique des Démosponges. Arch. Zool. Exp. Gén., 93: 1-181.

_____1957 Ontogeny and systematics in sponges. Syst. Zool., 6: 174-183.

_____1963 Gastrulation and larval phylogeny in sponges. In: The lower metazoa. (E.C. Dougherty et al., eds.),Univ. Calif. Press.

Little, F.J., Jr. 1963 The sponge fauna of the St. George's Sound, Apalachee Bay, and Panama city regions of the Florida Gulf coast. Tulane stud Zool., 11: 31-71.

Steedman, H.F. 1960 Section cutting in microscopy. Blackwell Sci. Pub., Oxford.

Topsent, E. 1893 Nouvelles séries de diagnoses d'éponges de Roscoff et de Banyuls. Arch. Zool. Exp. Gén., 1: 33-43.

_____1901 Spongiaires de l'expédition antarctique Belge. Résultats du voyage du S.Y. Belgica, 1894-1897, pp. 1-54.

Tuzet, O. and M. Pavans de Ceccatty 1955 La mobilisation en amoebocytes des cellules des _Halisarca_ (Éponges siliceuses). Les polyblastes chez les éponges. C.R. Soc. Biol. Paris., 149: 799-801.

_____1958 La spermatogénese, l'ovogénese, la fécondation et les premiers stade du développement d'Hippospongia communis LMK (H. equina O.S.). Bull. Biol. Fr. Belge, 92: 331–348.

Tuzet, O., R. Garrone, and M. Pavans de Ceccatty 1970 Origine choanocytaire de la lignée germinale mâle chez le Démosponge Aplysilla rosea Schulze (Dendroceratides). C.R. Acad. Sci. Paris, 270: 955–957.

Tuzet, O. and J. Paris 1964 La spermatogénese, l'ovogénése, la fécondation et les premiers stades du développement chez Octavella galangaui. Vie Milieu, 15: 309–327.

PLATES

PLATE 1

2. Sperm of <u>Halisarca</u> <u>dujardini</u> (A to G) and <u>H</u>. <u>nahantensis</u>
    (H to K). A to D are drawn from living sperm to show the
    disc-shaped head. E to G, drawn from smear preparations,
    do not show complete flagella. H and I are drawn from
    living sperm; J and K, from smear preparations.

3. Phase contrast photomicrograph of spermatozoa of <u>H</u>.
    <u>dujardini</u>.

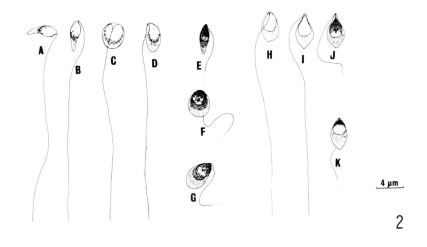

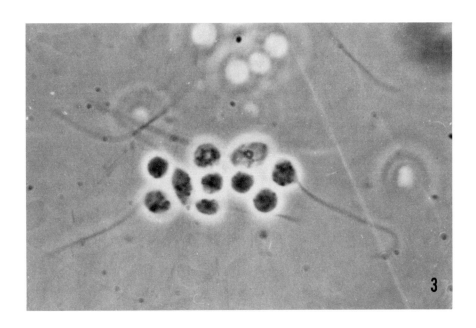

PLATE 2

The morphological changes in <u>Halisarca</u> <u>nahantensis</u> throughout
the reproductive period. Zenker's, Polyester, and Heidenhain's Azan.

4. <u>H</u>. <u>nahantensis</u> collected on 3/25/1973 showing the oocytes
    undergoing vitellogenesis. The number of "nucleolate"
    cells in the mesohyl is higher than that of the nonre-
    productive sponge. cc, choanocyte chamber; ex, exhalant
    canal; oo, oocyte.

5. <u>H</u>. <u>nahantensis</u> collected on 4/11/1973 showing vasculari-
    zation of the mesohyl. A similar phenomenon also occurs
    in <u>H</u>. <u>dujardini</u> during the same time of the year. cc,
    choanocyte chambers; ec, ectosome; oo, oocytes; vm, vas-
    cularization of the mesohyl.

6. <u>H</u>. <u>nahantensis</u> collected on 4/16/1973 showing the oocytes,
    spermatogonial aggregates, spermatic cysts and the rem-
    nants of choanocyte chambers. The aquiferous system is
    absent in this stage. ec, ectosome; cc, choanocyte
    chamber; oo, oocyte; sc, spermatic cyst; sg, spermatogo-
    nial aggregate.

7. Section of <u>H</u>. <u>nahantensis</u> during larval release. The
    choanocytes and the "nucleolate" cells are rounded and
    contain numerous cytoplasmic inclusions. One larva is
    leaving the nurse cavity into the exhalant canal. al,
    anterior end of larva; cc, choanocyte chamber; ex, ex-
    halant canal; fc, fuchsinophil cells; nc, nurse cavity;
    pl, posterior end of larva.

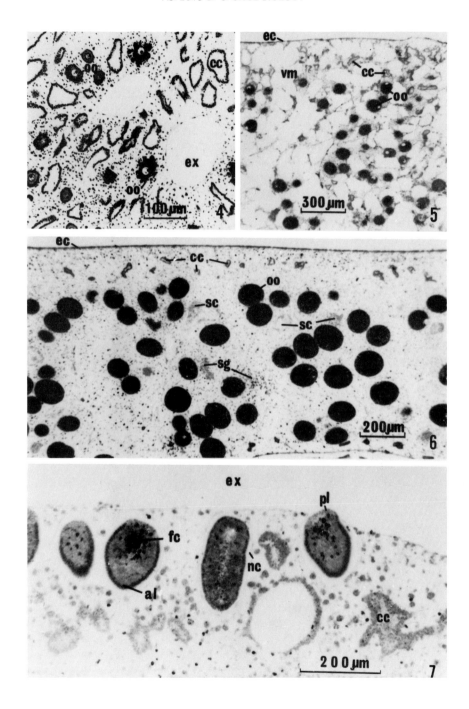

PLATE 3

The morphological changes in the same individual of <u>Halisarca</u> <u>dujardini</u> throughout the period of spermatogenesis. Zenker's, Polyester, and Heidenhain's Azan.

8. <u>H</u>. <u>dujardini</u> collected on 11/18/1972 showing the increase in number of cells in the mesohyl. The exhalant
canal and the choanocyte chambers are decreasing in
size. cc, choanocyte chamber; ex, exhalant canal; sg,
spermatogonia.

9. <u>H</u>. <u>dujardini</u> collected on 1/10/1973 showing the "nucleolate" cells and spermatogonia occupying the basal two
thirds of the mesohyl. The aquiferous canals have disappeared and the choanocyte chambers have closed. The
spermatogonia have aggregated to form a spermatic cyst.
cc, choanocyte chamber; ec, ectosome; sc, spermatic
cyst; sg, spermatogonial aggregate.

10. <u>H</u>. <u>dujardini</u> collected on 2/15/1973 showing many spermatic cysts in the basal portion of the sponge. All the
spermatic cysts contain mature spermatozoa. cc, choanocyte chamber; ec, ectosome; sc, spermatic cyst.

11. <u>H</u>. <u>dujardini</u> collected on 4/15/1973 showing vascularization of the mesohyl (vm).

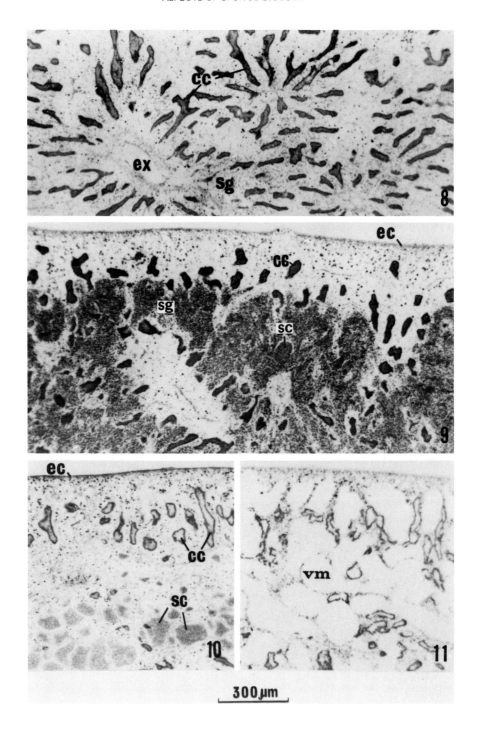

FEULGEN MICROSPECTROPHOTOMETRIC ANALYSIS
OF DEOXYRIBONUCLEOPROTEIN ORGANIZATION
IN SPONGE NUCLEI

Frederick W. Harrison
Department of Anatomy
Albany Medical College
Albany, New York   12208

and

Ronald R. Cowden
School of Medicine
East Tennessee State University
Johnson City, Tennessee   37601

ABSTRACT: Microspectrophotometric measurements of nuclei were undertaken on tissues from a variety of marine and freshwater sponges. Condensed nuclei of choanocytes and larval stage IV epitheliocytes of Eunapius fragilis and choanocytes of Ephydatia fluviatilis presented integrated extinction values about 64% and 85% respectively, of those obtained for either adult or larval archeocytes. This suggests that such measurements represent sensitive indicators of deoxyribonucleoprotein complex organization; thus, nuclear differentiation. Because of similarities of integrated extinction values of choanocytes and terminally differentiated stage IV larval epitheliocytes, it does not appear that choanocytes play a role in gametogenesis in freshwater sponges. In the absence of significant tetraploid, G_2, (4C), nuclear populations in any cell type, it is unlikely that a pool of tetraploid cells, available through mitotic increase for participation in morphogenetic events, is involved in regeneration in sponges. Integrated extinction values in choanocytes of freshwater sponges examined average 70% higher than integrated extinction values in choanocytes of the marine sponges examined. This increase possibly reflects increases in DNA content representing, in part, results of selection for duplication of genes whose products previously limited evolutionary adaptability – a concept applicable in the transition from the marine to a relatively unstable freshwater environment.

A considerable amount of literature is now available on the cytochemical demonstration and microspectrophotometric

141

measurement of DNA (Wied, '66; Wied and Bahr, '70). A number
of cytochemical methods directed toward either the demonstration of DNA or the protein portions of the deoxyribonucleoprotein (DNP) complex can also be used to detect differences
in nuclear DNP complex organization in cytological preparations (Ringertz, '69). As demonstrated by Garcia ('70) and
Cowden and Curtis ('75) the chromophore developed by the
Feulgen reaction, used under certain conditions, while
traditionally considered stoichiometric for DNA, may vary
considerably in G_1 cells of the same species. This is particularly true if high-resolution measuring systems are used.
These differences, apparently correlated with chromatin condensation, may in turn be related to the transcriptional
capacity of nuclear DNP. Variability in Feulgen chromophore
binding within G_1 cells from a given species appears in different cells at the various stages of differentiation within
a maturation series, or in similar cell types in different
metabolic states.

Because of these recognized variations in Feulgen chromophore related to DNP structure and function, it was necessary
to carry out interspecific determinations of the amount of
chromophore present in nuclei in a cell type that was differentiated and which presumably demonstrated a low level of
transcriptional activity. Choanocytes were selected as a
cell type that fit these criteria, and which was abundant in
tissue sections of all sponges used in this investigation.
By measuring Feulgen chromophore in choanocyte nuclei, it
was possible to initiate more extensive studies concerning
genome size evolution.

An examination of sections of the freshwater sponge,
Eunapius fragilis Leidy, indicated the presence of all stages
of larval development. The flagellated peripheral eiptheli-
ocytes of the spongillid parenchymula are a terminal cell
stage (Harrison and Cowden, 1975a), phagocytosed by underlying amebocytes following larval settling. The presence
of this cell type facilitated Feulgen microspectrophotometric
comparisons with choanocytes of the adult sponge to determine
similarities in chromophore content in nuclei of end-stage
differentiated cells of two distinct morphogenetic lines.

The presence of archeocytes in both larval and adult sponge
tissues allowed comparisons of the Feulgen chromophore produced in nuclei of the differentiated epitheliocytes and
choanocytes and the undifferentiated archeocytes. This system appeared to offer a particularly advantageous one in
which to examine the applicability of the Feulgen reaction
as a "probe of DNP organization".

As a corollary, the recent findings in coelenterates, based

upon both microspectrophotometric and autoradiographic investigations, of high proportions of tetraploid cells raised
the question of ploidy levels among sponge tissue cells. In
coelenterates, David and Campbell ('72) reported high levels
of tetraploidy in cells of Hydra attenuata while Curtis and
Cowden ('74) found a similar situation in Cassiopea sp. In
both cases, the tetraploid cells, apparently with 4C, G_2
nuclei, appeared to provide a cell pool which divided after
a traumatic event. Considering the high regenerative capacity exhibited by sponges, it seemed possible that an accumulation of cells with tetraploid G_2 nuclei might be present,
playing a similar role.

Our microspectrophotometric studies, then, have been focused on three areas: examination of the applicability of
the Feulgen reaction as a "probe of DNP organization"; examination of the possibility of the presence of a tetraploid
G_2 cell "pool" as a regenerative mechanism in sponges; and
obtaining integrated extinction values in a series of marine
and freshwater sponge choanocyte nuclei to initiate analysis
of the evolution of genome size in sponges.

MATERIALS AND METHODS

The "HI" series of specimens was collected at Heron Island,
Great Barrier Reef, Queensland, Australia. The spongillids,
Eunapius fragilis Leidy and Ephydatia fluviatilis (L.) were
collected from Six Mile Reservoir in Colonie, New York, a
suburb of Albany, New York and from City Park, New Orleans,
Louisiana. In all cases, tissues were fixed in ethanol-
acetic acid (3:1), routinely embedded in paraffin and sectioned at 5 µm. After de-waxing they were stained by the
Feulgen reaction using an 8-minute hydrolysis in 1N HCl.
Hydrolysis was stopped by immersion of the slides in ice
cold distilled water. Sections were then exposed to the
Schiff reagent for one hour. After three 10 minute washes in
SO_2 water, the preparations were dehydrated in a graded
ethanol series, cleared in xylol, and mounted in HRH
(Harleco, Philadelphia) mounting media.

These preparations were evaluated using the Zeiss "Cyto-
scan" system consisting of a Zeiss Universal stand, equipped
with a continuous interference filter wedge in the visible
range, and MPM-05 photometer head, a Zeiss photometric unit,
and an X-Y stepping stage. The entire system is interfaced
with a Digital Products PDP-12 computer using the May, 1972,
version of the APAMOS II program. This program allows pre-
selection and mapping of up to 63 cells, the scanning of a
field of variable size in 0.5 µm steps using a 0.5 µm

measuring spot, the integration of extinction values between preselected upper and lower limits, and the erasure by computer manipulation of adjacent cells or debris that might be in the measuring field. It reports in a teletype print-out the integrated extinction at a given wavelength (TE), the number of absorbing points between the preselected limits (TA), and the extinction per unit area (ME) for each nucleus. Only whole, uncut nuclei were measured. The accumulated microspectrophotometric data was evaluated by a variant of the Basic 69 statistical package for the "Student t-test" in general use with this computer.

OBSERVATIONS

Microscopic examination of adult archeocytes (Fig. 1) and choanocytes (Fig. 2) allows comparison of nuclear organization in an undifferentiated and a highly specialized cell type. Archeocyte nuclei are vesicular, containing considerable extra-chromosomal volume and a prominent nucleolus. Choanocytes contain condensed chromatin and no nucleoli. Nuclei of larval archeocytes (Fig. 3) are larger and more vesicular than nuclei of adult archeocytes. During morphogenesis of the epitheliocyte cell line, nuclei become progressively smaller and more condensed (Figs. 4 and 5). Mitosis may be seen in stage III epidermal cell nuclei (Fig. 4) while the condensed stage IV nuclei (Fig. 5) may assume a tear-drop shape. All details of larval development in Eunapius fragilis are presented in Harrison and Cowden ('75a).

Microspectrophotometric measurements indicated substantially higher mean integrated extinction (TE) values among archeocytes, adult and larval, than among choanocytes or stage IV epitheliocytes in both Eunapius fragilis and Ephydatia fluviatilis (Table 1). Both choanocytes and stage IV flagellated larval epitheliocytes of Eunapius fragilis presented mean TE values about 64% of those for either class of archeocytes. In Ephydatia fluviatilis choanocyte TE values were somewhat higher, about 85% of those for adult archeocytes. As table 1 illustrates, the larger more vesicular archeocyte nuclei bind more Feulgen chromophore than the smaller condensed nuclei of choanocytes or stage IV flagellated epitheliocytes. This relationship is seen in both Eunapius fragilis and Ephydatia fluviatilis. It is also apparent that the values obtained for archeocytes, larval and adult, represent diploid, 2C populations. A limited number of S and G_2 nuclei are also present in the larval archeocyte and stage III epitheliocyte populations of Eunapius fragilis (Fig. 6).

The TA (total absorbing points or total area, provided all

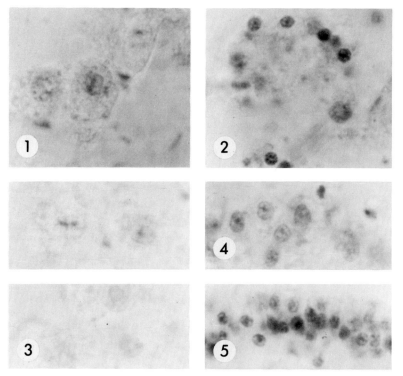

Fig. 1. Adult somatic archeocyte demonstrating vesicular nucleus with extended chromatin and large nucleolus. Feulgen method. 570 nm interference filter. X 2187.

Fig. 2. Adult choanocytes with nuclei containing condensed chromatin and no nucleoli. Feulgen method. 570 nm interference filter. X 2187.

Fig. 3. Larval archeocyte with nuclear organization similar to that seen in adult somatic archeocytes. Feulgen method. 570 nm interference filter. X 2187.

Fig. 4. Stage III larval epidermal cells exhibit an intermediate stage of chromatin condensation. Feulgen method. 570 nm interference filter. X 2187.

Fig. 5. Stage IV larval epidermal cells with highly condensed nuclear chromatin. Feulgen method. 570 nm interference filter. X 2187.

Figures 1-5 from Harrison and Cowden (1975b).

Table 1. Feulgen microspectrophotometric analysis of larval and adult freshwater sponge cell nuclei.[1]

Species	Cell Class	n	Integrated Extinction (TE)			Number of Absorbing Points (TA)		
			MEAN	SD	SE of MEAN	MEAN	SD	SE of MEAN
Eunapius fragilis	Adult Archeocytes	150	504.359	132.595	10.8263	61.5400	13.3479	1.08985
Eunapius fragilis	Larval Archeocytes	62	567.306	137.580	17.4727	62.5322	11.1070	1.41060
Eunapius fragilis	Choanocytes	102	350.529	84.9213	8.40847	17.5294	3.96694	.392786
Eunapius fragilis	Stage IV Epithelium	63	319.698	82.7880	10.4303	15.9206	3.45617	.435437
Ephydatia fluviatilis	Adult Archeocytes	155	472.006	14.0518	1.12867	81.86	3.14563	.252663
Ephydatia fluviatilis	Choanocytes	120	399.766	192.569	17.5791	20.0249	2.50986	2.29118

n, number of measurements

1 modified from Harrison and Cowden (1975b)

146

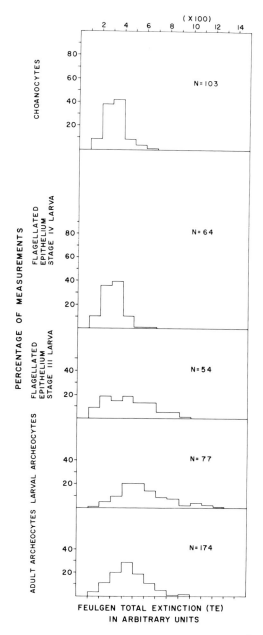

Fig. 6. Integrated extinction values (TE) for adult and larval nuclear populations (from Harrison and Cowden, 1975b).

147

sites within the structure register absorption within present
limits, which is the case) values are presented in histogram

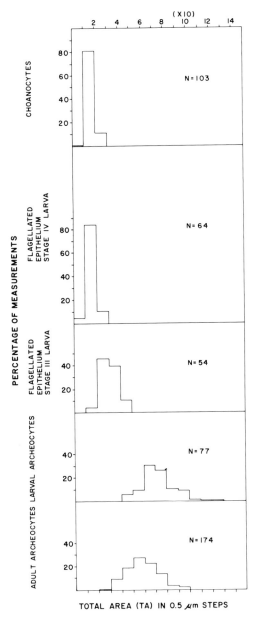

Fig. 7. Total absorbing points (TA) for adult and larval
nuclear populations (from Harrison and Cowden, 1975b).

form in figure 7. Archeocyte nuclei of Eunapius fragilis are considerably larger than nuclei of either differentiated cell type. Similarly, in Ephydatia fluviatilis, choanocytes present mean TA vaules about 25% of those for adult archeocytes (Table 1).

Preliminary comparisons of integrated extinction (TE) values in a series of marine sponge choanocyte nuclei with those obtained in Eunapius fragilis and Ephydatia fluviatilis choanocytes indicate much higher TE values in choanocyte nuclei of the freshwater sponges (Table 2). Composite choanocyte mean TE values in the marine sponges examined are only 30% of the composite choanocyte mean TE values of the two species. Choanocyte TE values in the two freshwater sponge species show a 13% difference.

Table 2. Means and standard deviations obtained
for total extinction (TE) and total absorbing points (TA)
in choanocyte nuclei of Feulgen-stained sponges

Species	n^1	TE^2	SD^3	SE of MEAN	TA^4	SD	SE of MEAN
Hymerhabdia sp.	30	110	47	8.5	10	2	0.4
Jaspis stellifera	30	82	48	8.8	6	2	0.4
Haliclona sp.	42	156	38	5.8	11	3	0.4
Haliclona sp.	33	76	36	6.2	7	2	0.3
Chondrilla australiensis	30	84	48	8.7	9	3	0.6
HI #30 (unidentified)	30	56	22	4.0	6	1	0.2
Callyspongia sp.	30	93	40	7.2	8	2	0.3
Spongia cf. nardorus	30	250	97	17.7	19	5	0.9
Ephydatia fluviatilis	120	400	193	17.6	20	3	2.3
Eunapius fragilis	103	351	85	9.2	18	4	0.4

[1]n, number of measurements

[2]TE, means of total extinction, to nearest whole number

[3]SD, standard deviation, to nearest whole number

[4]TA, mean of total number of absorbing points, to nearest whole number

[5]HI, Heron Island

DISCUSSION

Examination of morphological, morphometric and microspec-trophotometric data indicate that archeocytes exhibit the largest nuclei and bind the highest amounts of Feulgen chromophore. Cells with the most condensed nuclei, stage IV flagellated epitheliocytes, present a mean G_1 Feulgen chromophore value 56% of that seen in larval archeocytes, cells with the most vesicular nuclei.

Patterns of size, Feulgen chromophore content, and mean extinction were within expected limits (Ringertz, '69; Garcia, '70; Sawicki et al., '74; Cowden and Curtis, '75). The presence of both S and G_2 nuclei in larval archeocyte and stage III epitheliocyte nuclear populations indicates that deviations in Feulgen chromophore content between differentiated and undifferentiated cell classes reflect differences in chromatin organization and not differences in DNA content, i.e., ploidy. The presence of only a 15% difference in Feulgen chromophore values as TE in choanocytes and archeocytes of Ephydatia fluviatilis strongly supports this observation. As recently considered by Rasch and Rasch ('73), the observed differences may possibly be related to differential susceptibility to hydrolysis of DNA in different organizational relationships to protein conjugates. A possible relationship of these differences to transcriptional capacity of nuclear DNP would suggest that archeocyte DNP has high transcriptional capacity. Choanocyte DNP, exhibiting integrated extinction values (TE) similar to those of terminally differentiated larval epitheliocytes, has low transcriptional capacity. This is of some interest in view of reports of choanocyte roles in gametogenesis (see discussion in this volume). It seems unlikely that gametes, at least in the freshwater sponges examined, are derived from choanocytes.

There is no significant accumulation of tetraploid nuclei among archeocytes, nor any other class of cell nuclei, in either Eunapius fragilis or Ephydatia fluviatilis. The only nuclei which could be considered S phase or G_2 nuclei were a small number of larval archeocytes of stage III larval epitheliocytes of Eunapius fragilis. Both cell types were observed to display mitosis.

Archeocytes, a cell type almost identical with either blastomeres or larval archeocytes in nuclear morphology and DNP organization – and hence, transcriptional capacity, are present in considerable numbers in the adult sponge tissues. It therefore seems unlikely, in view of the abundance of this cell class with apparent blastomeric potential, that a special pool of tetraploid cells available for rapid increase of cells for participation in morphogenetic events, is required in the sponges.

Since choanocytes contain a 2C, G_1 nuclear population, mean integrated extinction (TE) vaules may be related to total DNA amounts (genome size), and offer information concerning the evolution of genome size in sponges. The pioneering work of Mirsky and Ris ('50) established evolutionary trends in DNA content per animal cell. There is now little doubt that genome size is subject to evolutionary pressures. The preliminary comparisons of TE and TA values of choanocyte nuclei from several Australo-Pacific species and freshwater sponges presented significant differences in DNA amounts per nucleus between marine and freshwater sponge nuclei. Although these apparent differences could conceivably reflect polyteny, polyploidy, or - less probably - aneuploidy, there is some possibility that the increase reflects differential gene redundancy. As Sparrow et al. ('71) have suggested, increases in DNA content may represent, in part, the results of selection for duplications of genes whose products previously limited evolutionary adaptability. This would certainly appear applicable in the transition from the marine to a relatively unstable freshwater environment.

ACKNOWLEDGMENTS

The "HI" series of sponges was collected while F.W. Harrison was Visiting Investigator, Heron Island Biological Station, Great Barrier Reef, Queensland, Australia. All Australo-Pacific specimens were kindly identified by Dr. Willard Hartman, Peabody Museum, Yale University and Dr. Patricia Bergquist, Department of Zoology, University of Aukland, New Zealand. Specimens of Ephydatia fluviatilis were collected and donated by Dr. Michael Poirrier, University of New Orleans, Louisiana.

LITERATURE CITED

Cowden, R.R. and S.K. Curtis 1975 Microspectrophotometric estimates of non-histone proteins in cell nuclei displaying differing degrees of chromatin condensation. J. Morph., 145: 1-12.

Curtis, S.K. and R.R. Cowden 1974 Some aspects of regeneration in the scyphistoma of Cassiopea (Class Scyphozoa) as revealed by the use of antimetabolites and microspectrophotometry. Am. Zool., 14: 851-866.

David, C.N. and R.D. Campbell 1972 Cell cycle kinetics and

development of Hydra attenuata. I. Epithelial cells. J.
Cell Sci., 11: 557–568.

Garcia, A.M. 1970 Stoichiometry of dye binding versus degree of chromatin coiling. In: Introduction to Quantitative Cytochemistry. II. G.L. Wied and F.G. Bahr, eds.
Academic Press, New York, pp. 153–170.

Harrison, F.W. and R.R. Cowden 1975a Cytochemical observations of larval development in Eunapius fragilis (Leidy):
Porifera; Spongillidae. J. Morph., 145: 125–142.

_____ 1975b Microspectrophotometric analysis of deoxyribonucleoprotein organization in larval and adult freshwater
sponge nuclei. J. Exp. Zool., 193: 131–136.

Mirsky, A.E. and H. Ris 1950 The desoxyribonucleic acid
content of animal cells and its evolutionary significance.
J. Gen. Physiol., 34: 451–462.

Rasch, R.W. and E.M. Rasch 1973 Kinetics of hydrolysis during the Feulgen reaction for deoxyribonucleic acid. A re-
revaluation. J. Histochem. Cytochem., 21: 1053–1065.

Ringertz, N.R. 1969 Cytochemical properties of nuclear
proteins and deoxyribonucleoprotein complexes in relation
to nuclear function. In: Handbook of Molecular Cytology.
A. Lima-de Faria, ed. American Elsevier Pub. Co., New
York, pp. 656–684.

Sawicki, W., J. Rowinski and J. Abramczuk 1974 Image analysis of chromatin in cells of preimplantation mouse embryos.
J. Cell Biol., 63: 227–233.

Sparrow, A.H., H.J. Price and A.G. Underbrink 1971 A survey
of DNA content per cell and per chromosome of prokaryotic
and eukaryotic organisms: some evolutionary considerations.
Brookhaven Symp., 23: 451–494.

Wied, G.L. 1966 Introduction to Quantitative Cytochemistry,
Academic Press, New York

Wied, G.L. and G.F. Bahr 1970 Introduction to Quantitative
Cytochemistry. II. Academic Press, New York.

THE EFFECTS OF THE CYTOCHALASINS ON SPONGE CELL REAGGREGATION: NEW INSIGHTS THROUGH THE SCANNING ELECTRON MICROSCOPE

Charlene Reed, Michael J. Greenberg and
Sidney K. Pierce, Jr.[1]

Department of Biological Sciences, Florida
State University, Tallahassee; and the Marine
Biological Laboratory, Woods Hole, Massachusetts

ABSTRACT: The surfaces of dissociated cells from the red sponge, Microciona prolifera, were examined with the scanning electron microscope. In normal sea water suspensions, a small amount of fibrous material is deposited on the cell surfaces. The administration of completely inhibitory concentrations of any of three cytochalasins-- A, B or E--increases the amount of this material. Moreover, the increment could be correlated with the potency of the particular cytochalasin as an inhibitor of reaggregation. Thus, cytochalasin E (CCE), the most potent inhibitor, also greatly augments the production of the fibrous material. Treatment with CCA is less effective; and CCB, the least potent inhibitor of the three drugs, followed. The surfaces of cells reaggregating in low calcium and magnesium sea water are smooth and have no fibrous material on their surfaces; but the addition of any of the cytochalasins increases the deposition of fibrous material to control levels. A working hypothesis is proposed suggesting at least two modes of action for the cytochalasins. One of these involves the production of excess aggregation factor, held to be identical with fibrous material. The excess factor, deposited on the cell surface, blocks binding sites, and thereby inhibits reaggregation.

Cytochalasins A, B and E inhibit the reaggregation of cells from dissociated Microciona prolifera, the red sponge (Greenberg & Pierce, '73; Greenberg et al., '76). However, we could observe both cytoplasmic streaming and pseudopod formation and withdrawal in sponge cells treated with concentrations of cytochalasin high enough to prevent any manifestation of reaggregation. Furthermore, the inhibition occurred whether the experiments were carried out in stationary dishes

---

[1]Current address: Department of Zoology, University of Maryland, College Park, Maryland

or on a rotating shaker table. Thus, no matter how the cells were making contact --- either by the motion of the table, or by their own movements --- they were not adhering once a contact was made. The evidence, in sum, strongly suggested that the inhibitory effect of the cytochalasins is not on cell motility. We were led to propose that the cytochalasins interfere with an intercellular adhesive process (Greenberg et al., '76).

The adhesive process in <u>Microciona</u>, as well as in other sponges, is mediated by specific aggregation factors produced and released by the dissociated cells (Moscona, '68; Humphreys, '63, '67, '70; McClay, '74). We therefore hypothesized that the cytochalasins could be affecting the production or release of the factor, or its binding to the cell surface (Greenberg et al., '76). Reaggregation of dissociated cells from embryonic chick cerebrum and retina is also augmented by specific factors (Garber & Moscona, '69; McClay & Moscona, '74). Moreover, various adhesive interactions between chick embryo cells can be modified or inhibited by cytochalasins B or A (Armstrong & Parenti, '72; Maslow & Mayhew, '72, '74; Steinberg & Weisman, '72; Sanger & Holtzer, '72; Appleton & Kemp, '74; Jones, '75). Thus, the proposed action of the cytochalasins in sponges might be extended to analogous systems in other phyla.

The actions of the three cytochalasins we have tested, CCA, CCB and CCE, are quantitatively and qualitatively different. First, in terms of the concentration necessary to arrest reaggregation permanently at any stage of the process, CCE was the most potent compound, CCA was next and CCB was least active. Second, the cytochalasins also produce diffuse, cloudy suspensions of cells which settle very slowly at high concentrations, attaching only weakly to the bottom of the dish. CCE was, again, most effective in eliciting this response, with CCA and CCB following in that order; but CCE was also much more active in preventing settling than would have been expected on the basis of its relative potency. Thus, CCE-treated suspensions remained cloudy for some hours, while completely inhibitory doses of CCB produced relatively little of this effect. These results cannot distinguish between possible mechanisms of cytochalasin action since, in theory, the effects could have been brought about by either an excess or an insufficiency of aggregation factor.

One way of determining whether the cytochalasins increase or decrease aggregation factor production -- if, indeed, they do either -- is to examine the surfaces of variously treated cells with the scanning electron microscope (SEM). The aggregation factor is a fibrous glycoprotein of about 21 x 10^5

daltons; and transmission electron micrographs reveal the fibrous components to be arranged in a loose "sunburst configuration" about 0.15-0.3 μm across (Henkart et al., '73; Cauldwell et al., '73). Thus, these molecules should be visible with the SEM. In addition, when cells are dissociated into calcium-magnesium-free sea water (CMF-SW), the factor is removed from their surfaces (Humphreys, '63, '67, '70). Therefore, we looked with the SEM for a fibrous material on sponge cell surfaces identifiable with the structure seen in the transmission electron microscope. We expected that it sould be absent from the cell surface in low calcium and magnesium medium, but present in control cells; and that it should vary, in some predictable way, with the relative inhibitory potencies of CCA, CCB and CCE on sponge cell reaggregation.

MATERIALS AND METHODS

We obtained <u>Microciona</u> <u>prolifera</u> from the Northeast Marine Specimens Company, Inc., Bourne, Massachusetts; they arrived in Tallahassee, Florida in good condition within 33 hours of shipment <u>via</u> U.S. Postal Service (Airmail/Special Delivery). The sponges were kept in aquaria, in natural sea water from the Gulf of Mexico (31 ppt), at 15°C, and were vigorously aerated. Even under these conditions, the animals finally began to degenerate in about three weeks; but we used animals only within the first week after their receipt.

The <u>Microciona</u> cell suspensions were prepared by carefully squeezing 10 grams of sponge through several layers of cheesecloth into 100 ml of sea water in a large fingerbowl. The cheesecloth was alternately squeezed and dipped in the sea water until a bright orange suspension was obtained. Aliquots (2.5 ml) of the suspensions were added to 7.5 ml of sea water in Syracuse dishes, yielding a final concentration between 4-8 x 10^7 cells per ml. To observe reaggregation in low Ca^{++}-Mg^{++} sea water (LCM-SW), 2.5 ml aliquots of a fresh sponge suspension were added to 7.5 ml of Ca^{++}-Mg^{++}-free sea water (Cavanaugh, '56), rather than to regular sea water.

Stock solutions (2 μg/ml) of the cytochalasins (Imperial Chemicals Industries, Ltd., Macclesfield, Cheshire, U.K.) were prepared in dimethylsulfoxide (DMSO); doses of the stock were added to the sea water in the Syracuse dishes with a microliter syringe to achieve the final concentration in the suspension. In all of the experiments reported here, doses of cytochalasin that completely inhibited reaggregation were employed (Greenberg, et al., '76): CCA and CCE,

0.5 µg/ml; and CCB, 7.0 µg/ml.

The preparation of sponge suspensions for scanning electron microscopy proceeded as follows. The sponge cell suspensions were prepared as usual; with or without cytochalasin; and in normal or LCM-SW. Gold-palladium-coated nucleopore filters (1 mm^2) were placed on the bottom of each dish, so that the aggregates or individual sponge cells would collect on them.

One hour after preparing the suspensions, the cells were fixed by adding 1% gluteraldehyde (320 mOsM; pH 7.4 in 0.1 M phosphate buffer) to each Syracuse dish. After 10 minutes, the filters with the fixed cells and aggregates still attached, were transferred to a fresh solution of 1% gluteraldehyde at 4°C, for at least 2 hours, but for as long as 2 days. Following the gluteraldehyde fixation, the tissues were dehydrated in a series of graded alcohols, followed by a set of graded amyl acetate concentrations (50%, 70%, 80%, 95% and 100%; 5 minutes in each concentration). The samples were then dried in CO_2 by the critical point method. The dried material on the nucleopore filters was attached to the SEM stubs with double-sided adhesive tape, and was coated with approximately 100 Å of gold-palladium using a vacuum evaporator (Denton model 502). The specimens were observed with a scanning electron microscope (Cambridge S4-10) operated at 10 to 20 KV.

RESULTS

Untreated cells

Observed with the scanning electron microscope, the surfaces of sponge cells fixed in normal sea water are rough and randomly ornamented with small blebs of about 0.6 to 0.8 µm in diameter. A few strands of fibrous material deposited on the cell surface are also visible. The fibrils are 0.1 µm in diameter (Figs. 1a and 1b).

Effect of the cytochalasins

Treatment with any of the cytochalasins alters the appearance of the cell surfaces in that the amount of fibrous material is increased. The extent of this increase is dependent on the particular cytochalasin used. The surfaces of cells treated with CCE were almost obscured by 0.1 µm fibrils (Figs. 1g and 1h). Fibrous material also covers most of the surface of the CCA-treated cells, although not to as great an extent as those exposed to CCE (Figs. 1c and 1d). The surfaces of CCB-treated cells are least affected; yet there are more fibers deposited on them than are on the

surfaces of control cells (compare Figs. 1e and 1f with 1a).
In summary, CCE treatment caused the greatest augmentation
of this material; less was found after CCA treatment, and
still less after CCB. The amount of fibrous material deposited by each cytochalasin is therefore correlated with its
potency as an inhibitor of reaggregation.

Effect of the suspension medium on reaggregation

Cell suspensions were treated with 0.5 μg/ml of CCE and,
after 60 minutes, were centrifuged at 3,000 rpm for 2 minutes
to remove the cells. The supernatant was passed through an
ultrafiltration membrane (Amicon PM-10) which retains molecules larger than 10,000 daltons. This procedure removed
the cytochalasins from the cell suspensions. The material
retained on the filter was taken up in 15 ml of sea water,
and this was added to a freshly prepared suspension of dissociated Microciona cells in normal sea water. The effect
of this treatment was to accelerate the appearance of the
early stages of reaggregation in comparison with control suspensions (Table 1).

TABLE 1

Reaggregation rate: the effect of the medium from
a suspension treated with cytochalasin E (CCE)

Stage[1]	Experimental[2]	Control[3]
I	27.5	42.5
II	49.0	52.0
III	59.2	67.0
IV	83.5	89.0

Data are mean times (min) to stages I-IV of reaggregation.
[1]Stages of reaggregation: As the numerals increase, the
 clumps of aggregated cells become larger, denser, more
 spherical, and smoother in outline (see Greenberg et al.,
 1976).
[2]Experimental suspension: 2.5 ml of CCE-treated suspension
 medium added; CCE removed by ultrafiltration.
[3]Control suspension: Reaggregation occurring in normal sea
 water.

The effect of low calcium and magnesium
The surfaces of sponge cells suspended in LCM-SW were
noticeably different from those of the controls: the surfaces

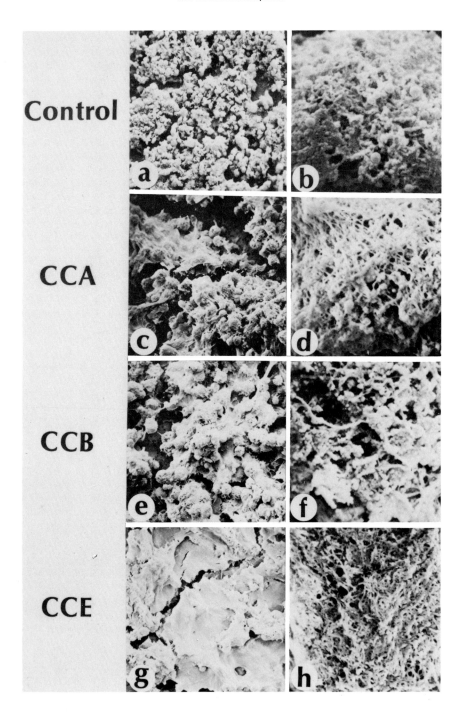

Control

CCA

CCB

CCE

a

b

c

d

e

f

g

h

Fig. 1. Scanning electron micrographs showing the effects
of completely inhibitory doses of the cytochalasins on the
surfaces of dissociated cells of Microciona prolifera.Left-
hand column: low magnification, showing many cells in suspension; right-hand column: high magnification, showing the
surfaces of a single cell. Control (no cytochalasin): (a)
x680; (b) x16,000. CCA (0.5 μg/ml): (c) Increased amount
of fibrous material in the suspension, x720; (d) fibrous
material covers most of the cell surface, x15,000. CCB
(7 μg/ml): Some increase in the quantity of fibrous material
over controls; but not as great as is seen with the other
cytochalasins; (e) x720; (f) x15,000. CCE (0.5 μg/ml): (g)
Note the vast quantity of fibrous material present in the
suspension, x720; (h) a cell surface completely obscured by
fibrous material, x15,000.

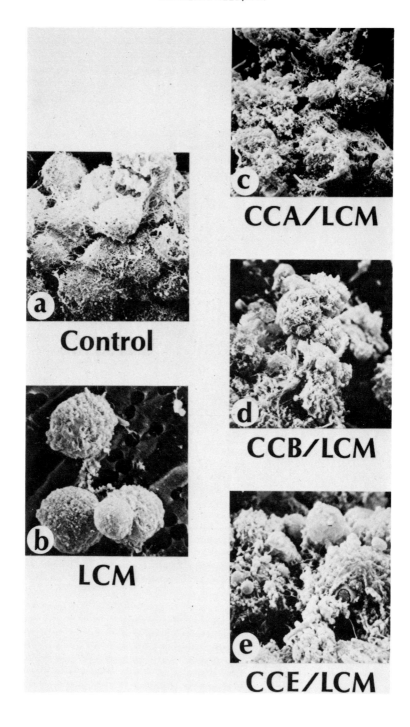

160

Fig. 2. Scanning electron micrographs showing the effect of
low calcium and magnesium (LCM) sea water, alone and with
added cytochalasins, on the surfaces of dissociated cells
from Microciona prolifera. The concentrations of cytochalasins used completely inhibited reaggregation. (a) Control:
the cells are reaggreagating in normal sea water. x3,500.
(b) LCM sea water: no reaggregation occurring; cell surfaces
are relatively smooth and no fibrous material is present.
x3,100. (c) CCA (0.5 μg/ml) in LCM sea water. x3,100. (d)
CCB (7 μg/ml) in LCM sea water. x3,500. (3) CCE (0.5 μg/ml)
in LCM sea water. x3,000. Note that all of the cytochalasins increase the amount of fibrous material to control
levels.

were smoother; there were fewer blebs; and, most important, no fibrous material was present (compare Figs. 2a and 2b). Addition of any of the three cytochalasins to cell suspensions in LCM-SW caused the cell surfaces to resemble those of the controls again: the number of blebs increased and the fibrous material appeared on the surfaces (Fig. 2). No reaggregation occurred in LCM-SW suspensions treated with cytochalasin.

DISCUSSION

Scanning electron microscopy of reaggregating cells from the red sponge, Microciona prolifera, together with earlier findings (Greenberg et al., '76), have led us to a preliminary explanation of the mechanism of action of cytochalasins in this system. The working hypothesis can be stated in four points.

i) The fibrous material normally observed on the surfaces of cells, and enhanced when the cells are treated with cytochalasins, is the aggregation factor proposed by Humphreys ('63) and by Moscona ('63).

ii) A corollary is that at least one effect of the cytochalasins on reaggregating sponge cells is to augment the production of aggregation factor.

The identity between the fibrous material and the aggregation factor is supported, firstly, by the parallelism between the relative potencies of the three cytochalasins on the reaggregation process and on the production of the fibrous material. Thus, the compound most effectively evoking the increase in deposited fibrous material is CCE; CCA is less potent and CCB is least active. This ranking was found although completely inhibitory concentrations were employed, with a resulting dose ratio:

$$[CCE] : [CCA] : [CCB] = 1 : 1 : 12$$

favoring a larger response from CCB-treated cells. The relative effectiveness of the cytochalasins in inhibiting the reaggregation of mechanically dissociated sponge cells can be similarly ranked: CCE > CCA > CCB. Moreover, the same order of potency obtains when the production of diffuse, viscid, slowly settling cell suspensions is considered (Greenberg et al., '76).

Secondly, the surfaces of cells suspended in a medium low in calcium and magnesium are devoid of deposited fibrous material. Such cells also lack factor since, according to the classic experiment of Humphreys ('63), they will not reaggregate in normal sea water at $5^{o}C$ unless the dissociation medium is added back to the suspension. This further parallelism between morphology and function also supports the identity of the fibrous material and aggregation factor. Moreover, when any of the cytochalasins are added to the LCM suspensions, fibrous material can once again be observed on the cell surfaces to the extent of the control cells in normal sea water. Of course, cytochalasin-treated LCM suspensions do not reaggregate; in the low divalent ion concentration of the medium, the factor is presumably dissociating as it is produced (Humphreys, '73; Cauldwell et al., '73). We are able to see factor on the surfaces of LCM-suspended cells only because the cytochalasins have increased the rate of its production.

Finally, given the differences in the methods of preparation and observation, the chemical and morphological description of the aggregation factor as a large, fibrous proteoglycan (Henkart et al., '73; Cauldwell et al., '73) is sufficiently suggestive of the fibrous material seen in the SEM so as to support the identity between the two.

iii) The action of aggregation factor is biphasic. Although low, normal surface concentrations promote reaggregation, and small increments of the factor may further enhance the rate of the process, a sufficiently large excess bound to the cell surface actually inhibits reaggregation. Thus, the production of excess factor by the cytochalasins inhibits sponge cell reaggregation.

Probably, aggregation factor binds to specific sites on the cell surface and thus promotes adhesion of like cells upon contact (see the model of Weinbaum and Burger, '73). Additional factor would then increase the rate at which the bonds were made, as well as the strength of the total interaction. However, if an excess of factor were produced, all the receptor sites would be occupied and no cell-cell binding would take place. Such a mechanism is analogous, for example, to the well-known requirement for optimal proportions of antigen and antibody in immune precipitation reactions.

The elaboration of a layer of excess surface-bound aggregation factor can explain the production, by CCE and CCA, of viscid clouds of suspended cells. Presumably, the coating of excess factor prevents direct cell-cell adhesion in the suspension, while weak interactions between the most periferal of the layered factor molecules result in the diffuse, slowly settling clouds. In sufficient excess (i.e., in high doses of CCE), factor might form a gel throughout the suspension, halting settling altogether. In fact, we have observed such gels by performing reaggregation experiments in 10 ml graduated cylinders in the presence of completely inhibitory doses of CCE. Moreover, gelation in sea water by high concentrations of aggregation factor has been reported by Henkart et al. ('73).

The concept of blockade by high concentrations of factor would seem, at first, to be in conflict with the clear-cut experiments of Humphreys ('63) and McClay ('74). They added excess homospecific factor to reaggregating cells with the result that extraordinarily large aggregates were produced (see, especially, Fig. 7a in Humphreys, '63). However, while these workers necessarily dissolved aggregation factor in the medium, the effect of cytochalasin is to increase production of factor: thus, the excess is deposited in place, on the cell surface, and before bridging between cells can occur.

Nevertheless, if CCE is producing excess aggregation factor, then, under appropriate circumstances, an <u>increase</u> in cell adhesion should be manifested. Some evidence that this occurs is available. Firstly, when completely inhibited CCE-treated suspensions are washed out, cell adhesion rebounds: aggregation occurs more rapidly, and to a greater extent, than in the untreated but washed controls (Fig. 3). Presumably, the wash has not only removed the cytochalasin, but has also reduced sufficiently the amount of aggregation factor bound to the surface so that binding sites are unmasked and rapid adhesion can take place. Secondly, this possibility is supported by the finding that the suspending medium of CCE-treated cells from which the cytochalasin has been washed by ultrafiltration can accelerate the early stages of reaggregation when added to a fresh, untreated suspension.

Finally, in the initial 15 minutes of its action, a high dose of CCE actually increases the rate of settling of sponge cells from suspension (Greenberg et al., '76). An analogous response seems to occur in chick embryo cells. Aggregation of unmixed or homotypic cells is variably inhibited by cytochalasin B, and, in some cases, is unaffected (see references in Introduction); but in two instances, aggregation of such cells was augmented. First, Jones ('75) found that, in 0.5 µg/ml CCB, reaggregation of neural retina cells was "consistently a little more rapid within the first 15 minutes than was that of the control cells". Second, the reaggregation of chick heart ventricle cells is dramatically enhanced by CCB, a single, large, irregular aggregate being formed by the end of 1 day (Steinberg & Wiseman, '72; see Fig. 5).

iv) <u>There must be at least one additional inhibitory mechanism of cytochalasin action which does not involve the production of excess aggregation factor.</u>

The primary datum supporting this proposition is that CCB

produces little excess fibrous material (aggregation factor) even in high concentrations that completely inhibit reaggregation of dissociated sponge cells. Furthermore, only the action of another inhibitory mechanism can explain the absence of enhancement of aggregation at judiciously chosen low concentrations of CCA and CCE. On the other hand, the high potency of these compounds, compared with CCB, might follow from the action of both mechanisms at moderate to high doses. Finally, accepting the risk of reasoning from sponges to fowl, the variable effect of cytochalasins on the formation of homotypic chick embryo cell aggregates could result from the interplay of two such mechanisms.

The nature of the second inhibitory action proposed here remains obscure. Presumably, however, the cytochalasins could be attaching to either of the complementary active sites -- on the aggregation factor itself, or the cell surface -- preventing binding and thus adhesion.

The possibility that cytochalasins may be acting at binding sites on the factor or the cell surface, and at sites triggering the production of factor, has an additional implication. Assuming the structural complementarity usually existing between drugs and their receptors, our hypothesis suggests that the binding sites and the factor production sites are similar, and that, in fact, the amount of aggregation factor bound to a cell surface may regulate the rate at which the cell produces additional factor.

ACKNOWLEDGMENTS

The authors thank Mr. Bill Miller for his expert assistance with the scanning electron microscopy and the preparation of figures. This is Contribution Number 47 from the Tallahassee, Sopchoppy and Gulf Coast Marine Biological Association. Support was provided by the National Institutes of Health (grant HL-09283) and by Sigma Xi.

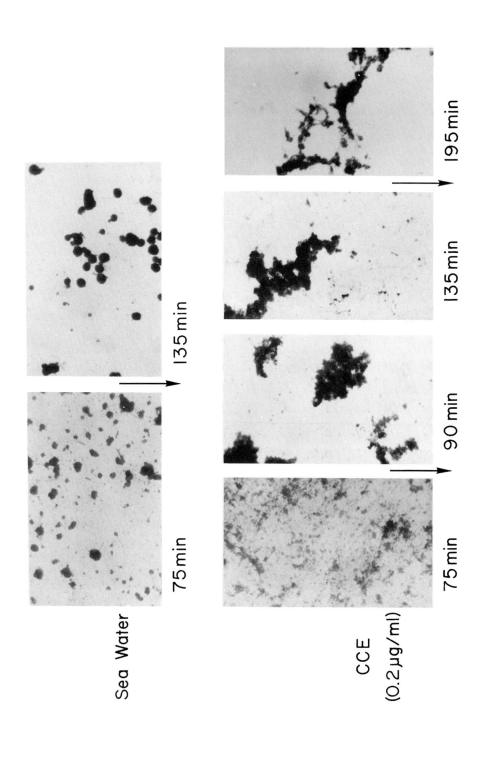

Sea Water

75 min   135 min

CCE
(0.2 µg/ml)

75 min   90 min   135 min   195 min

Fig. 3. Suspensions of dissociated cells from <u>Microciona</u> <u>prolifera</u>: effects of changing the suspending medium. The initial dissociation was at 0 min; photographs were taken at various times after dissociation, as indicated. At the arrows (75 min; 135 min) the medium was replaced with sea water. Upper photographs (control suspension in sea water); washing has no effect on reaggregation; compact, smooth-surfaced spheres form by 135 min. Lower photographs (CCE-inhibited suspensions): large, irregular masses form after the first wash and persist after the second.

LITERATURE CITED

Appleton, J.C. and R.B. Kemp 1974 Effects of cytochalasins on the initial aggregation *in vitro* of embryonic chick cells. J. Cell Sci., 14: 187-196.

Armstrong, P.B. and D. Parenti 1972 Cell sorting in the presence of cytochalasin B. J. Cell Biol., 55: 542-553.

Cauldwell, C.B., P. Henkart and T. Humphreys 1973 Physical properties of sponge aggregation factor. Biochemistry, 12: 3051-3055.

Cavanaugh, G.M. (Ed.) 1956 *Formulae and Methods V. of the Marine Biological Laboratory Chemical Room.* Marine Biological Laboratory, Woods Hole. p. 71.

Garber, B.B. and A.A. Moscona 1969 Enhancement of aggregation of embryonic brain cells by extracellular materials from cultures of brain cells. J. Cell Biol., 43: 41a.

Greenberg, M.J. and S.K. Pierce, Jr. 1973 A new look at sponge cell reaggregation: Inhibition by the cytochalasins. Amer. Zoologist, 13: 1336-1337.

Greenberg, M.J., C. Reed, and S.K. Pierce, Jr. 1976 Dissociated cells of *Microciona prolifera* (Porifera) are inhibited from reaggregating by cytochalasins A, B, and E. J. Exp. Zool. (in press)

Henkart, P., S. Humphreys and T. Humphreys 1973 Characterization of sponge aggregation factor. A unique proteoglycan complex. Biochemistry, 12: 3045-3050.

Humphreys, T. 1963 Chemical dissolution and *in vitro* reconstruction of sponge cell adhesions. I. Isolation and functional demonstration of components involved. Develop. Biol., 8: 27-47.

_____ 1967 The cell surface and specific cell aggregation. In: *The Specificity of Cell Surfaces*, B.D. Davis and L. Warren, eds. Prentice-Hall, Inglewood Cliffs, N.J., pp. 195-210.

_____ 1970 Biochemical analysis of sponge cell aggregation. In: *The Biology of the Porifera*, W.G. Fry, ed. Symp. Zool. Soc. Lond. No. 25: 325-334.

Jones, G.E. 1975 Effects of cytochalasin B on the aggregation, electrophoretic mobility and surface morphology of chick neural retina cells. J. Cell Sci., 17: 371-379.

Maslow, D.E. and E. Mayhew 1972 Cytochalasin B prevents specific sorting of reaggregating embryonic cells. Science, 177: 281-282.

_____ 1974 Histotypic cell aggregation in the presence of cytochalasin B. J. Cell Sci., 16: 651-663.

McClay, D.R. 1974 Cell aggregation: Properties of cell surface factors from five species of sponge. J. Exp.

Zool., 188: 89–102.

McClay, D.R. and A.A. Moscona 1974 Purification of the specific cell-aggregating factor from embryonic neural retina cells. Exp. Cell Res., 87: 438–443.

Moscona, A.A. 1968 Cell aggregation: Properties of specific cell-ligands and their role in the formation of multicellular systems. Develop. Biol., 18: 250–277.

Sanger, J.W. and H. Holtzer 1973 Cytochalasin B: Effects on cell morphology, cell adhesion and mucopolysaccharide synthesis. Proc. Nat. Acad. Sci. U.S.A., 69: 253–257.

Steinberg, M.S. and L.L. Wiseman 1972 Do morphogenetic tissue rearrangements require active cell movements? The reversible inhibition of cell sorting and tissue spreading by cytochalasin B. J. Cell Biol., 55: 606–615.

Weinbaum, G. and M.M. Burger 1973 Two component system for surface guided reassociation of animal cells. Nature, 224: 510–512.

CYTOLOGICAL ABNORMALITIES IN A VARIETY OF NORMAL AND TRANSFORMED CELL LINES TREATED WITH EXTRACTS FROM SPONGES[1]

Jack T. Cecil, Martin F. Stempien, Jr.,
George D. Ruggieri, and Ross F. Nigrelli

Osborn Laboratories of Marine Sciences,
New York Aquarium, New York Zoological Society,
Brooklyn, New York   11224

ABSTRACT: Aqueous extracts from dried West Indian
sponges, Teichaxinella morchella, Cribrochalina sp.,
Adocia carbonaria, Agelas sp., Verongia archeri, and
Xestospongia muta contain biologically active compounds
which induce cytological abnormalities in KB cells
(human oral carcinoma), GSK cells (transformed gray
seal kidney), and the fish cell lines FHM (fat head
minnow), and RTG-2 (rainbow trout gonadal).   None of
these extracts have induced cytological abnormalities
in WI-38 cells (normal human embryonic lung).   Extracts
from dried West Indian sponges Plakortis sp., Haliclona
erina, Haliclona rubens, and Haliclona sp. do not produce cytological abnormalities when used in the same
concentrations as the above sponges, in any cell line
thus far tested.   Cells were either passed and grown
in dilutions containing 25-50 micrograms/ml. of the
extracts, or the cells were grown first as monolayers,
and extracts applied in similar dilutions in the
maintenance medium.   Observations were made by fluorescent microscopy of acridine orange stained materials,
by phase microscopy, or by hematoxylin-eosin stained
preparations.   Abnormalities include nuclear fragmentation, formation of micronuclei in "rings", nuclear and
nucleolar budding, and giant cell formation.

A number of toxic and pharmacologically active substances
have been obtained from marine invertebrates.   Cohen ('66);
Doering et al., ('66); Leopold ('65); and Underwood et al.,
('65) have shown that D-arabinosyl nucleosides demonstrated
biological effects in the treatment of several virus infections such as Herpes simplex keratitis and leukemias in man
and laboratory animals.   These substances were first isolated from the West Indian sponge Crypotethya crypta in

---

[1]This research was supported by a grant from the Scaife
Family Charitable Trusts; also from Sea Grant - U.S. Department of Commerce.

large quantities. Bergmann and Burke ('56); and Bergmann and Stempien ('57) were the first to chemically characterize these substances as the 1-B-D arabinofuranosyl derivatives of thymine (spongothymidine) and uracil (spongouridine). It was pointed out by Cohen ('63) that these substances could be of possible therapeutic value.

Stempien et al., ('69) demonstrated that extracts from several sponges possess antibiotic and antimycotic activity, produce ichthyotoxins, possess several biological activities on developing sea urchin eggs, and also show nuclear, cyto-plasmic and cytotoxic effects on KB human oral carcinoma cell lines. Nigrelli et al., ('67) have shown that sponges con-tain antibiotic and antitumor substances. Tan et al., ('73) have demonstrated cytotoxic effects from alcoholic extracts of the sponge Suberites inconstans, on the human cervical carcinoma cell line, designated HeLa.

In our continuing research on the cytopathology of bio-logically active compounds from marine invertebrates, we have recently focused our attention to extracts from marine sponges, and their effects on a variety of normal and trans-formed cell lines in vitro. In this report we consider: 1) the cell culture assay system, 2) preparation of the ex-tracts, 3) description of activity among the different species of marine sponges, 4) description of the multiple micronuclear "ring" formation, nuclear and nucleolar "bud-ding" and giant cell formation, 5) possible mechanisms which may be involved in these anomalies.

MATERIALS AND METHODS

A. Cell cultures:

Transformed cell culture lines have been established from a number of animal species (Willmer '67). In our research we have studied cytological effects, caused by extracts from marine sponges, on the continuous cell line KB, which was derived from a human oral carcinoma (Eagle '62); and the transformed cell line from the kidney of a gray seal, charac-terized by the letters GSK (Cecil and Nigrelli '70). This cell line has been in continuous passage in our laboratory for 75 generations, at which time the cell line was frozen. This research was achieved between the 55th and 65th pas-sages.

Additional cell culture lines were assayed for cytological abnormalities against sponge extracts. These cell lines in-cluded: 1) a cell line designated FHM, fat head minnow, which was established by Gravel and Malsberger ('65); this cell line has a chromosome modal distribution of 51, 2n=50,

2) the cell line RTG-2, was established by Wolf and Quimby ('62) from pooled gonadal tissue of male and female rainbow trout; this cell line has a modal chromosome number ±2 of 59, 2n=60, 3) the normal human embryonic cell line from the lung, WI-38, was established by Hayflick and Moorhead ('61). All cell lines were grown in minimum essential medium of Eagle ('62), supplemented with 10 per cent fetal bovine serum.

B. Sponge extracts:
  Marine sponges (Table I) were collected from the coastal waters surrounding both Jamaica and the British Virgin Islands during the past five years. The sponges were oven dried at a temperature not exceeding 80°C for a period of 48 hours. Portions of each of these samples were ground in a mortar and then the powdered samples were stored in screw cap sealed glass tubes at 4°C. Activity was apparently not diminished by this storage procedure.
  A 100 mgm sample of each sponge was then placed in 10 ml. of cellular growth medium, stirred vigorously for 10 min., centrifuged at 1000 rpm. for 10 min., and the supernatant layer containing the dissolved extract, saved. In order to determine the amount of dissolved extract in the solution, a 0.1 ml. aliquot was evaporated onto a weighed standard peni- cillin assay disk, and its weight compared with a similar disk prepared from 0.1 ml. medium. Dilutions of extracts were made to contain a final concentration of 25-50 μgm/ml. in minimum essential growth medium.

C. Visualization of activity:
  Dilutions of the water-soluble portions of the sponge extracts were prepared w/v, in the medium described above, and sterilized through a swinney millipore filter. Cells were removed from stock and 5 x 10^4 cells/ml. inoculated in- to Leighton tubes. These cells were then grown to monolayer. The old medium was discarded from the cells, and new medium, containing the extracts, placed on the cells. Other dissoci- ated cells were placed in growth medium containing the ex- tracts, and 5 x 10^4 cells/ml. inoculated into Leighton tubes. Leighton tube cultures of cells and extracts, and control cells were incubated at 37°C for mammalian cells or 20°C for fish cells.
  At intervals after treatment with sponge extracts, cover- slip preparations were stained with acridine orange fluoro- chrome and observed by fluorescent microscopy. Other cover slip preparations were observed by phase microscopy or stained with hematoxylin and eosin.

D. Acridine orange staining:

This staining procedure offers the property of cytochemical differentiation (Armstrong '56). DNA-components fluoresce yellow-green while RNA-components fluoresce rust-red. Fresh coverslip preparations were rinsed in buffer, then placed in a solution of acridine orange (1:1000) in buffer, and stained for 15 minutes. Coverslip preparations were rinsed in two changes of buffer. Coverslips were mounted in buffer for fluorescent microscopic examination.

E. Fluorescent microscopy:

An American Optical Fluorolume microscope and illumination system were used. An Osram HBO 200 watt bulb in conjunction with a BG-12 exciter filter and an OG-6 barrier filter was the illumination source. A Zeiss 35mm camera was used, with a Zeiss automatic timer. Either type B Kodak Ektachrome, Panatomic-X, or Tri-X film was used.

F. Phase microscopy:

Phase was utilized either by cover slips on slides observed by American Optical phase equipment, or by the use of a Wild #60 inverted microscope and phase objectives.

RESULTS

A. Relative activities of sponge extracts:

Samples of sponge extracts were tested using minimum essential medium, with Earle's balanced salt solution, supplemented with non-essential amino acids and fetal bovine serum. Medium, containing the extracts, was placed on the mono-layered cell cultures, or extracts were incorporated in the medium from stock cultures containing 5×10^4 cells/ml. and inoculated into Leighton tubes. The additional milliosmolarity between control and extract containing medium was negligible.

Depending upon the extent of morphological abnormalities and cytotoxicity produced, activity was recorded on a comparative basis as + = 25%, ++ = 50%, +++ = 75%, ++++ = 100%. Results are recorded in Table I.

Sponges used in this research are shown in Figures 1 through 10.

B. KB cells are illustrated in Figure 11.

C. Nuclear fragmentation and a giant cell containing four nuclei are shown in Figure 12.

D. Nuclear and nucleolar budding in the transformed cell lines KB and GSK are pictured in Figures 13 and 14.

E. Micronuclear "ring" formation in transformed cells is illustrated in Figure 15.

TABLE I

RELATIVE ACTIVITIES OF SPONGE EXTRACTS

Sponge extract 25-50 μgm/ml.	Cell lines				
	KB	GSK	FHM	RTG-2	WI-38
Teichaxinella morchella	+++	+++	+	+	−
Chribrochalina sp.	+++	+++	+	+	−
Adocia carbonaria	+++	+++	+	+	−
Agelas sp. (JC-6)	++++	++++	++	++	−
Verongia archeri (hard)	++++	++++	++	++	−
Xestospongia muta	++	++	+	+	−
Plakortis sp. (JC-44)	−	−	−	−	−
Haliclona erina	−	−	−	−	−
Haliclona rubens	−	−	−	−	−
Haliclona sp. (CI-207)	−	−	−	−	−

Fig. 1. Teichaxinella morchella   Fig. 2. Xestospongia muta

Fig. 3. <u>Chribrochalina</u> sp.

(JC-20)

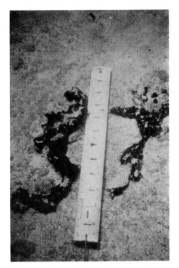

Fig. 4. <u>Adocia</u> <u>carbonaria</u>

Fig. 5. <u>Agelas</u> sp.

(JC-6)

Fig. 6. <u>Verongia</u> <u>archeri</u>

(hard form)

Fig. 7. Plakortis sp.

(JC-44)

Fig. 8. Haliclona erina

Fig. 9. Haliclona rubens

Fig. 10. Haliclona sp.

(CI-207)

Fig. 11. KB cells

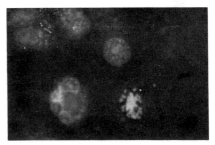

Fig. 12. Nuclear frag-
mentation in KB cells

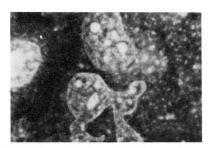

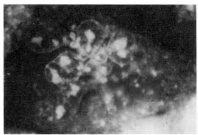

Fig. 13. Nuclear budding in transformed cells

Fig. 14. Nucleolar budding in transformed cells

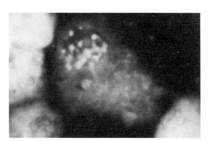

Fig. 15. Micronuclear ring formation in transformed cells

DISCUSSION

In this paper we describe some of the nuclear and cytoplasmic abnormalities resulting after exposure of normal and transformed cell culture lines to extracts from marine sponges.

Exposure of transformed cell lines to extracts from the sponges Teichaxinella morchella, Cribrochalina sp., Verongia archeri, and Xestospongia muta results in striking nuclear and cytoplasmic anomalies. These abnormalities include nuclear fragmentation, formation of micronuclei in "rings", nuclear and nucleolar budding, and giant cell formation. Exposure of non-transformed fish cell lines to these extracts results in a much lower percent of cells containing nuclear and cytoplasmic anomalies. The normal human embryonic lung cell line WI-38 apparently is not affected by these sponge extracts. Extracts from the sponges Plakortis sp., Haliclona erina, Haliclona rubens, and Haliclona sp., do not produce cytological abnormalities, when used in the same concentrations as the cytotoxic-producing sponges, in any cell lines thus far tested. The occurrence of all these conditions in the preparations could imply: 1) a concentration effect, 2) multiple activity of a single compound, 3) the presence of several biologically active substances.

Starr et al., ('66) have obtained micronucleation, amitosis, and multiple cytokinesis or miniature cell formation, utilizing marine macro-algal extracts on the McCoy synovial fluid cell line. He further postulated some of these anomalies may be due to multiple spindle formation. Tan et al., ('73) have shown cytoplasmic fragmentation and prominent nuclear membrance formation in HeLa cells, utilizing extracts from the sponge, Suberites inconstans.

In this report we have observed nuclear fragmentation and micronuclear "ring" formation. As Starr et al., ('66) suggest, this could possibly be due to multiple, rather than single, spindle operative mechanisms. We have not, however, observed amitosis in our treatments with sponge extracts. Another possibility that exists for the controlling mechanism of nuclear fragmentation, micronuclear "ring" formation, giant cell formation and also nuclear and nucleolar budding is that of a selective nuclear membrane active substance. This substance evidently is not effective against the cytoplasmic membrane but is active in causing nuclear abnormalities. Further experimentation on additional transformed and non-transformed cell lines, exposed to other sponge extracts, should provide additional basic biological information relating to processes involved in these anomalies.

LITERATURE CITED

Armstrong, J.A.   1956   Histochemical differentiation of
nucleic acids by means of induced fluorescence.   Exp. Cell
Res., 11: 640-643.

Bergmann, W. and D.C. Burke   1956   Contributions to the study
of marine products XXXIV.   The nucleosides of sponges III.
Spongothymidine and spongouridine.   J. Org. Chem., 20:1501.

Bergmann, W. and M.F. Stempien, Jr.   1957   Contributions to
the study of marine products XLIII.   The nucleosides of
sponges V.   The synthesis of spongosine.   J. Org. Chem.,
22: 1575.

Cecil, J.R. and R.F. Nigrelli   1970   Cell cultures from
marine mammals.   J. Wildl. Dis., 6: 494-495.

Cohen, S.S.   1963   Sponges, cancer chemotherapy and cellular
aging.   Perspectives Biol. Med., 6: 215.

_____1966   Introduction to the biochemistry of D-arabinosyl
nucleosides.   In: Progress in Nucleic Acid Research. Vol.
V.   (J.N. Davidson and W.E. Cohn, eds.) Academic Press,
New York.   pp. 1-88.

Doering, A.J., J. Keller and S.S. Cohen   1966   Some effects
of D-arabinosyl nucleosides on polymer synthesis in mouse
fibroblasts.   Cancer Res., 26: 2444.

Eagle, H.   1955   Propagation in a fluid medium of a human
epidermoid carcinoma, strain KB (21811).   Proc. Soc. Biol.
(N.Y.), 89: 362.

Gravel, M. and R.G. Malsberger   1965   A permanent cell line
from the fathead minnow (Pimephales promelas).   Ann. N.Y.
Acad. Sci., 126: 555-565.

Hayflick, L. and P.S. Moorhead   1961   The serial cultivation
of human diploid strains.   Exp. Cell Res., 25: 585-621.

Leopold, I.H.   1965   Clinical experience with nucleosides in
Herpes simplex eye infections in man and animals.   Ann.
N.Y. Acad. Sci., 130: 181.

Nigrelli, R.F., M.F. Stempien, Jr., G.D. Ruggieri, V.R.
Liguori and J.R. Cecil   1967   Substances of potential bio-
medical importance from marine organisms.   Fed. Proc., 26:
1197-1205.

Starr, R.J., M. Kajima and M. Piferrer   1966   Mitotic
anomalies in tissue cultured cells treated with extracts
derived from marine algae.   Tex. Rep. Biol. Med., 24: 208-
221.

Stempien, M.F., Jr., G.D. Ruggieri, S.J., R.F. Nigrelli and
J.T. Cecil   1970   Physiologically active substances from
extracts of marine sponges.   In: Food-Drugs from the Sea
Proceedings 1969 (H.W. Yongken, Jr., ed.) Marine Tech-

nology Society, Washington, D.C. pp. 295-305.

Underwood, G.E., G.S. Elliott and D.A. Buthala 1965 Herpes keratitis in rabbits. Pathogenesis and effects of anti-viral nucleosides. Ann. N.Y. Acad. Sci., 130: 151.

Willmer, E.N. 1965 Morphological problems of cell type, shape, and identification. In: Cells and Tissues in Culture. Vol. I. (E.N. Willmer, ed.) Academic Press, N.Y. pp. 143-176.

Wolf, K. and M.C. Quimby 1962 Established eurythermic line of fish cells in vitro. Science, 135: 1065-1066.

ARE DEMOSPONGIAE MEMBRANES UNIQUE AMONG LIVING ORGANISMS?

Carter Litchfield and Reginald W. Morales

Department of Biochemistry
Rutgers University
New Brunswick, New Jersey   08903

ABSTRACT: Demospongiae membrane lipids contain high levels of 24 to 30 carbon fatty acids instead of the usual 14 to 22 carbon acids found in other organisms. The extra bulk of these longer fatty acid chains probably produces a distinctive lipid bilayer structure in Demospongiae membranes and may alter some of their physical and physiological properties.

Membranes are a much studied subject today; we all know their importance. They constitute the boundaries between living and nonliving matter and between various compartments of the cell. The selective metabolic processes occurring at the surface of all these membranes are critical factors for the maintenance of life.

There are many ways to examine membranes; most of these methods fall into three basic approaches. First of all, one can examine membrane morphology. For this purpose, the many elegant techniques of electron microscopy have proven quite effective. A second approach to membrane examination is to break them apart and then define the molecular building blocks that they are made of. Numerous new biochemical analysis techniques now permit us to make such compositional studies in considerable detail. A third approach to understanding membranes is to define the dynamics of mass flow across them: the transport of various metabolites into and out of the cell, for example. Here, unfortunately, our experimental techniques are not equal to the task. We can only take a crude "black box" type approach to defining what goes in and what comes out. At present we have no adequate methods to define the microarchitecture of the membrane itself so that we can fully understand how such transport is accomplished.

This paper will concern itself with the building block or molecular analysis approach to the characterization of membranes within the class Demospongiae of the phylum Porifera. Our recent experiments have shown that many of the lipid membrane components found in Demospongiae cells have distinctly different structures from those found in other

living organisms. If these membrane building blocks are so
distinctive, then it follows that Demospongiae membranes
could be quite different from those of other organisms.

MEMBRANE FATTY ACIDS

The familiar fluid mosaic model for membranes (Fig. 1) has
been described by Singer et al. ('72). It consists of a
planar lipid bilayer (i.e. two molecules thick) in which
numerous proteins are embedded. This lipid bilayer is con-
structed from large numbers of phospholipid and sterol molecules oriented with their hydrophylic polar groups (the circles in Fig. 1) directed outward and the hydrophobic hydrocarbon chains (the wiggly lines in Fig. 1) in the interior.

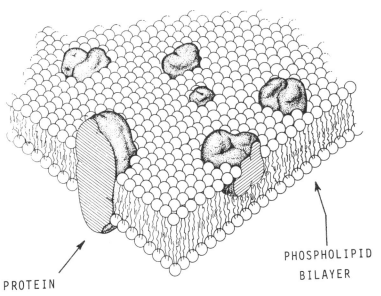

PHOSPHOLIPID
BILAYER

PROTEIN

Fig. 1. Schematic, cross-sectional view of the fluid mosaic
model for membranes. It consists of a planar lipid bilayer,
two phospholipid molecules thick, in which numerous proteins
are embedded (Singer et al., '72). Copyright 1972 by the
American Association for the Advancement of Science.

Most of this hydrocarbon interior consists of fatty acid
chains from the phospholipid molecules. The lipid bilayer
is called a "fluid mosaic" because its molecular components

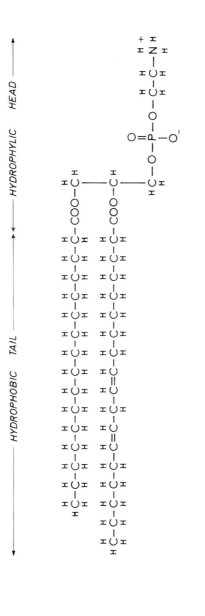

Fig. 2. Structural formula for phosphatidylethanolamine, a typical membrane phospholipid. Other types of phospholipids are also found in the membrane lipid bilayer (Gurr & James, '71); and they all contain a similar hydrophobic tail at one end of the molecule and a hydrophylic head at the other.

are in constant motion. The lipid molecules move about rapidly within their half of the lipid bilayer, but there is no appreciable exchange between the inside and outside halves of the bilayer. Membrane-bound proteins also move about in this fluid mosaic, perhaps maintaining a fixed domain of specific lipids around them. Some proteins are found only on the outside or inside surface of the membrane, while others extend completely through the bilayer.

The complete structural formula for phosphatidylethanolamine, a common phospholipid molecule found in membranes, is shown in figure 2. The hydrophobic/hydrophylic dichotomy of this molecule is clearly evident here. The water-insoluble fatty acid chains both lie at one end, while the water-soluble portions (glycerol, phosphate, ethanolamine) are all located at the other end. In an aqueous environment, such phospholipid molecules will cluster together with their hydrophylic ends pointing outwards toward the water phase, and the hydrophobic fatty acid chains pointing inward. The resultant structure is like the lipid bilayer found in natural membranes. Comparison of figures 1 and 2 shows that the longer the fatty acid chains, the thicker the lipid bilayer will be.

It is the hydrophobic interior of the cell membrane that constitutes the major barrier between the intra- and extracellular aqueous solutions of metabolites. Hydrophylic molecules such as water, mineral ions, sugars, amino acids, etc. entering or leaving the cell must somehow penetrate this hydrophobic barrier. It has been shown that the physical properties of the fatty acids found in the membrane phospholipids have a great deal to do with membrane permeability. The rates at which various molecules pass through the lipid bilayer are highly dependent on the fatty acid composition of the phospholipids (De Gier et al., '68; Haest et al., '69).

Typical fatty acids found in membrane phospholipids are shown in figure 3. They possess straight chains 14 to 22 carbon atoms long with an acid carboxyl group (-COOH) at one end. This carboxyl is connected to the polar head group, usually glycerol, in the membrane phospholipid molecules. From zero to six double bonds are found along the fatty acid chain, usually with a 1,4-relationship to each other (i.e. $-\underset{1}{C}=\underset{2}{C}-\underset{3}{C}-\underset{4}{C}=\underset{5}{C}-$) when more than one is present. For convenience in naming these acids, the following shorthand nomenclature has been widely adopted: (number of carbon atoms): (number of double bonds). Thus 18:1 is oleic acid having a chain 18 carbons long with one double bond. Physical properties of fatty acids, such as viscosity, melting point,

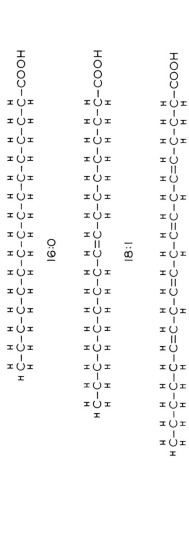

Fig. 3. Typical fatty acid chains found in the membranes of plants, animals, and microorganisms.

etc. are directly related to the number of carbon atoms and to the number of double bonds present in the long chain. Decreasing the chain length or increasing the number of double bonds makes liquid fatty acids less viscous and the membrane that contains them more fluid. Increasing the chain length or decreasing the number of double bonds in membrane fatty acids makes that membrane more rigid.

If one draws a matrix giving all possible combinations of carbon atoms and double bonds using the shorthand nomenclature and then compares this matrix with the fatty acids usually found in the membranes of living organisms, a very regular pattern is observed. Figure 4 presents such a matrix with an indication of the major ($\geq$5%) acids found in the membranes of all levels of organisms except the Porifera (Table 1). One immediately notes that these major membrane fatty acids all fall into a central region of the matrix. We have drawn a heavy black line around this group of membrane acids and called it the "region of feasible adaptations."

Fig. 4. Matrix of possible membrane fatty acids classified according to the number of carbon atoms (number before the colon) and the number of double bonds (number after the colon). Major ($\geq$5.0%) fatty acids occurring in the membranes of plants, animals (except the Porifera), and microorganisms all fall within the "region of feasible adaptations". Data sources listed in Table I.

TABLE I

SOURCES CONSULTED TO DETERMINE THE MAJOR FATTY ACIDS FOUND IN THE
MEMBRANE LIPIDS OF LIVING ORGANISMS[a]

Type of Organism	Genus & Species	Sample Analyzed	Reference
wheat	Triticum vulgare	seed phospholipids	McKillican & Sims ('64)
potato	Solanum tuberosum	tuber mitochondria lipids	Abdelkader et al. ('69)
spinach	Spinacea oleracea	leaf lipids	Debuch ('61)
spruce tree	Picea sp.	leaf lipids	Schlenk & Gellerman ('65)
cycad	Cycas revoluta	leaf lipids	Schlenk & Gellerman ('65)
fern	Matteucia struthiopteris	leaf lipids	Schlenk & Gellerman ('65)
horsetail	Equisetum arvense	total lipids	Schlenk & Gellerman ('65)
club moss	Lycopodium sp.	total lipids	Schlenk & Gellerman ('65)
peat moss	Sphagnum sp.	"leafy part" lipids	Schlenk & Gellerman ('65)
green alga	Halosphaera viridis	total phospholipids	Ackman et al. ('70a)
diatom	Nitzschia closterium	total lipids	Kates & Volcani ('66)
brown alga	Fucus serratus	total lipids	Klenk et al. ('63)
red alga	Rhodomelia subfusca	total lipids	Klenk et al. ('63)
blue-green alga	Spirulina platensis	total polar lipids	Nichols & Wood ('68)
yeast	Saccharomyces cerevisiae	plasma membrane lipids	Longley et al. ('68)
bacterium	Streptococcus pyogenes	membrane lipids	Cohen & Panos ('66)
slime mold	Dictyostelium discoideum	total phospholipids	Davidoff & Korn ('63)
phytoflagellate	Euglena gracilis	total lipids	Korn ('64)
zooflagellate	Crithidia sp.	total phospholipids	Korn et al. ('65)

(Table I – continued)

Type of Organism	Genus & Species	Sample Analyzed	References
ameba	Acanthamoeba castellani	plasma membrane lipids	Ulsamer et al. ('71)
jellyfish	Cyanca capillata	total polar lipids	Sipos & Ackman ('68)
sea anemone	Metridium dianthus	total phospholipids	Hooper & Ackman ('71)
mollusk	Ostrea edulis	total polar lipids	Watanabe & Ackman ('74)
insect	Culex pipiens fatigans	total phospholipids	Subrahmanyam et al. ('71)
crustacean	Meganystiphanes norvegica	total phospholipids	Ackman et al. ('70b)
turtle	Dermochelys coriacea	lung polar lipids	Ackman et al. ('72)
fish	Ammodytes americanus	total phospholipids	Ackman & Eaton ('71)
seal	Phoca vitulina	heart polar lipids	Ackman et al. ('72)
rat	Rattus rattus	liver plasma membrane lipids	Ray et al. ('69)
rat	Rattus rattus	intestinal microvillus plasma membrane lipids	Forstner et al. ('68)
man	Homo sapiens	red blood cell phospholipids	Dodge & Phillips ('67)

aFor this compilation, a major membrane fatty acid is defined as 5.0% or more of the fatty acids in the total polar lipids of a tissue. Representative data on typical membrane structures from as many phyla as possible have been included. No data on nerve tissue are included because its distinctive myelin membrane structure has not been found in the Demospongiae. Where separate polar lipid analyses were not available, a total lipid analysis has been substituted only if the tissue analyzed probably contains no major neutral lipid component with fatty acid moieties (brown algae, bacteria, plant leaves, etc.). In the unusual cases where odd chain length fatty acids are present, they have been combined with the next higher even homologue (i.e. 17:1 + 18:1) to simplify presentation.

The fatty acid composition of a living membrane is not
fixed but can vary with environmental conditions such as
temperature (Haest et al., '69). Such adaptations in poikilothermic organisms point to a control mechanism which main-
tains the cell membrane at a desirable permeability so that
transport across that membrane will be normal even if environmental parameters change. Lowering the number of carbon
atoms or increasing the number of double bonds in the fatty
acid chains would make the lipid bilayer more fluid. Increasing the number of carbon atoms or lowering the number
of double bonds will make the lipid bilayer more rigid. The
control mechanism regulating membrane permeability apparently
works mainly with those fatty acids located within the
"region of feasible adaptations". Shorter chain acids, for
example, would make the bilayer too fluid so that it "leaks".
Dolphin fat cells are known to contain high levels of 5:0
(isovaleric acid), but none of that 5:0 is found in the membrane lipids of these cells (Varanasi & Malins, '74). Similarly, longer, less unsaturated acids apparently make the
bilayer too rigid and nonpermeable. Although 34% 22:1 is
present in the triglycerides (energy storage lipids) of
rapeseed, the 22:1 is excluded from the membrane lipids of
those seeds (Appelqvist, '75).
Membrane biochemists have generally considered this "region
of feasible adaptations" to be universal among all eucariotic
and procariotic organisms. But now the Demospongiae appear
to be an exception to this rule.

DEMOSPONGIAE FATTY ACIDS

Twenty-five years ago Bergmann and Swift ('51) reported
the presence of unusually high levels of 24, 26, and 28
carbon fatty acids in the tissues of two Demospongiae, Speci-
ospongia vesparia and Suberites compacta. After reading this
paper, curiosity led us to examine the fatty acids of the
red beard sponge, Microciona prolifera, which is readily
available in New Jersey coastal waters. To our surprise, we
found large amounts of 26 carbon fatty acids in Microciona
(Morales & Litchfield, '76). The two major long chain acids
were identified as 26:2 and 26:3 (Jefferts et al., '74) having the molecular structures shown in figure 5.
Since fatty acids normally occur as parts of larger lipid
molecules in living organisms, our next step was to determine
which class of lipids these 26 carbon acids came from. Were
these acids combined into energy storage lipids such as
triglycerides or wax esters, or did they appear in membrane

191

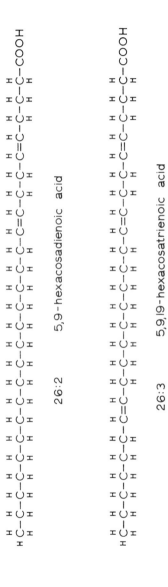

Fig. 5. Molecular structures of the unusual 26 carbon fatty acids found in the membrane lipids of *Microciona prolifera* (Jefferts et al., '74).

192

components such as phospholipids? Our investigations showed
that Microciona contains four major classes of lipids: sterol
esters, phosphatidylethanolamine, phosphatidylcholine, and
phosphatidylserine (Table II). These last three phospholipids have structures of the type shown in figure 2 and are
typical membrane components found in many organisms. The
exact cellular location and function of sterol esters in
living tissues is still uncertain. When the fatty acid com-
positions of the individual lipid classes were examined,
very high levels of the 26 carbon acids were found in the two
membrane lipids phosphatidylethanolamine and phosphatidylserine (Table II).

TABLE II

DISTRIBUTION OF 26 CARBON FATTY ACIDS IN MICROCIONA LIPIDS

	Fatty Acids (wt. %)	
	26:1 + 26:2	26:3
TOTAL LIPIDS	14	31
Phosphatidylserine	22	48
Phosphatidylcholine	1	3
Phosphatidylethanolamine	12	29
Neutral lipids (mainly sterol esters)	35	23

Further study of the 26:1, 26:2, and 26:3 fatty acid content of Microciona tissues through the May-to-December annual
growth cycle (Figure 6) revealed that 26:1 and 26:2 were much
higher in the summer months while 26:3 predominated in the
fall months. This variation correlates well with the measured water temperature in a manner similar to other growth temperature studies of poikilothermic organisms (Haest et al.,
'69). Membrane fatty acids with more double bonds are required at lower temperatures to keep membrane flexibility at
its optimum level. Conversely, membrane fatty acids contain
fewer double bonds at higher temperatures to prevent the
membrane lipid bilayer from becoming too fluid. Thus the
seasonal variations observed for 26:1 , 26:2, and 26:3 content
provide further evidence for the presence of these acids in
Microciona membranes.

We are now in the midst of a study to determine if high
levels of C_{24}-C_{30} fatty acids are widespread among the class
Demospongiae or merely limited to a few families or orders.
Our results to date combined with those of Bergmann & Swift

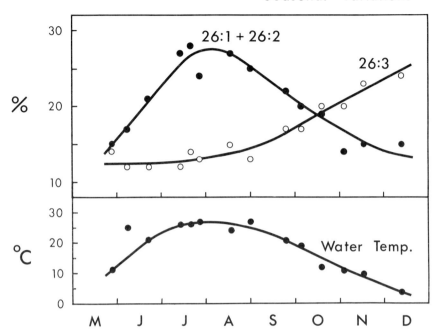

Fig. 6. Variation of 26:1 + 26:2 and 26:3 content of <u>Micro-ciona prolifera</u> fatty acids and of water temperature during the May-to-December annual growth cycle for this animal.

are summarized in Table III. Eleven genera representing five orders within both the Ceractinomorpha and Tetractino-morpha have been examined to date, and all contain similar high levels of these extremely long fatty acid chains. No exceptions have been found so far. Preliminary examination of the original lipid mixtures of many of these sponges by thin-layer chromatography showed mainly membrane-type polar lipid classes like <u>Microciona</u>. Our tentative conclusion is that C_{24}-C_{30} membrane fatty acids are widespread, possibly ubiquitous, within the entire class Demospongiae.

DEMOSPONGIAE MEMBRANES

If high levels of C_{24}-C_{30} fatty acids are characteristic of the membrane lipids found in Demospongiae tissues, what effect will these unusual components have on the physical properties of those membranes? In figure 7, data on the

TABLE III

DISTRIBUTION OF FATTY ACID CHAIN LENGTHS
IN DEMOSPONGIAE TISSUES

Order	Genus & Species	Fatty Acids (Wt.%)		Reference
		$C_{14}-C_{22}$	$C_{23}-C_{30}$	
Astrophorida	Chondrilla nucula	55	45	a
Hadromerida	Anthosigmella varians	67	33	a
	Cliona celata	21	79	a
	Speciospongia vesparia	42	58	b
	Suberites compacta	50	50	b
Halichondrina	Halichondria panicea	35	65	a
Haplosclerida	Haliclona oculata	36	64	a
Poesilosclerida	Isotrochota birotulata	36	64	a
	Lissodendoryx sp.	38	62	a
	Microciona prolifera	49	51	c
	Tedania ignis	39	61	a

a. Litchfield et al., '75.
b. Bergmann & Swift, '51.
c. Morales & Litchfield, '76.

fatty acid composition of Microciona prolifera membrane
lipids have been inserted into the fatty acid matrix developed in figure 3. Note that the three C_{26} Microciona fatty
acids fall outside the "region of feasible adaptations" defined for organisms outside the Porifera. As figure 7 indicates, these high levels of 26:1, 26:2, and 26:3 should make
Microciona membranes entirely too rigid to maintain normal
levels of membrane-associated metabolic reactions. From
experience with other organisms, one would expect cell at-
tenuation, atrophy, or death to occur with such acids in the
membrane lipids. But obviously the membranes of Microciona,
as well as the entire class Demospongiae, do operate quite
efficiently. Therefore, we infer that there is something

195

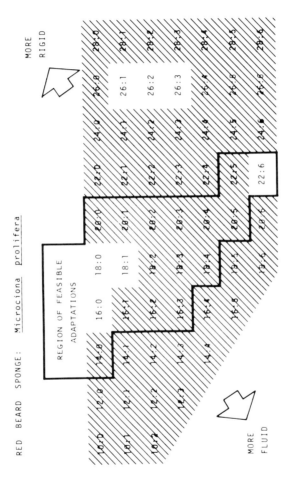

RED BEARD SPONGE: _Microciona prolifera_

Fig. 7. Major (≥5%) membrane fatty acids of _Microciona prolifera_ (Morales & Litchfield, '76) inserted into the membrane fatty acid matrix developed in figure 3. Note that the 26 carbon fatty acids of _Microciona_ fall outside the "region of feasible adaptations" defined for organisms other than the Porifera.

196

unusual about their structure and/or mode of operation.

Returning to the fluid mosaic model for membranes (Fig. 1), how would these extremely long $C_{24}-C_{30}$ fatty acids affect its structure? Obviously there must be some modification to accommodate the extra carbon atoms in the fatty acid chains. Perhaps the lipid bilayer becomes 6 to 12 carbon atoms thicker, depending on whether the longer acid chains are found on one or both sides of the membrane. Perhaps the $C_{24}-C_{30}$ fatty acid chains may fold or twist in some manner in order to be accommodated in the lipid bilayer. In such a case, the surface area occupied by individual phospholipid molecules would be greater and components of the lipid bilayer would be spaced further apart. Unusual phospholipid-sterol interactions might also take place.

The exact modification of Demospongiae membranes by the high levels of $C_{24}-C_{30}$ fatty acids remains to be elucidated. Nevertheless, it seems fairly certain that some modification of the usual membrane structure must exist because of the sheer bulk of the longer fatty acid chains inserted. We conclude, therefore, that Demospongiae membranes probably have a distinctive lipid bilayer structure not found in other living organisms. This would make Demospongiae cells useful model systems for studying the effects of fatty acid chain length on membrane parameters.

This conclusion certainly raises a major physiological question. Just why did the Demospongiae evolve membranes having such high levels of $C_{24}-C_{30}$ fatty acids? What special function would these unusual fatty acids have? We would welcome speculative hypotheses from any who read this article. Such ideas could help guide us in designing future experiments.

ACKNOWLEDGMENTS

This investigation was supported in part by grants from the Rutgers Marine Sciences Center and the Rutgers University Research Council. R.W. Morales gratefully acknowledges receipt of a Johnson & Johnson Fellowship in Biology.

LITERATURE CITED

Abdelkader, A.B., P. Mazliak and A.M. Catesson 1969 Biogenese des lipides mitochondriaux au cours de la "survie" (ageing) de disques de parenchyme de tubercule de pomme de terre. Phytochemistry, 8: 1121-1133.

Ackman, R.G., R.F. Addison, S.N. Hooper and A. Prakash 1970a *Halosphaera* viridis: fatty acid composition and

taxonomical relationships. J. Fish. Res. Bd. Canada, 27:
251-255. (The fatty acid composition of Halosphaera
phospholipids can be estimated by subtracting the fatty
acid composition of the triglyceride fraction from that of
the total lipids.)

Ackman, R.G. and C.A. Eaton 1971 Investigation of the fatty
acid composition of oils and lipids from the sand lance
(Ammodytes americanus) from Nova Scotia waters. J. Fish.
Res. Bd. Canada, 28: 601-606.

Ackman, R.G., C.A. Eaton, J.C. Sipos, S.N. Hooper and J.D.
Castell 1970b Lipids and fatty acids of two species of
North Atlantic krill (Meganyctiphanes norvegica and Thy-
sanoëssa inermis) and their role in the aquatic food web.
J. Fish. Res. Bd. Canada, 27: 513-533.

Ackman, R.G., S.N. Hooper and J. Hingley 1972 The harbor
seal Phoca vitulina concolor De Kay: comparative details of
fatty acids in lung and heart phospholipids and triglycerides. Can. J. Biochem., 50: 833-838.

Appelqvist, L.-Å. 1975 Biochemical and structural aspects
of storage and membrane lipids in developing oil seeds.
In "Recent Advances in the Chemistry and Biochemistry of
Plant Lipids", T. Galliard and E.I. Mercer, eds., Academic
Press, London, p. 253.

Bergmann, W. and A.N. Swift 1951 Marine products.XXX.
Component acids of lipides of sponges. J. Org. Chem., 16:
1206-1221.

Cohen, M. and C. Panos 1966 Membrane lipid composition of
Streptococcus pyrogenes and derived L-form. Biochemistry,
5: 2385-2392.

Davidoff, F. and E.D. Korn 1963 Fatty acids and phospholipid composition of the slime mold Dictyostelium discoide-
um. J. Biol. Chem., 238: 3199-3209.

Dabuch, H. 1961 Über die fettsauren aus grünen blättern
und das vorkommen der Δ³-trans-hexadecensäure. Z. Naturforsch., 16b: 561-567.

DeGier, J., J.G. Mandersloot, and L.L.M. van Deenen 1968
Lipid composition and permeability of liposomes. Biochim.
Biophys. Acta, 150: 666-675.

Dodge, J.T. and G.B. Phillips 1967 Composition of phospholipids and of phospholipid fatty acids and aldehydes in
human red cells. J. Lipid Res., 8: 667-675.

Forstner, G.G., K. Tanaka and K.J. Isselbacher 1968 Lipid
composition of the isolated rat intestinal microvillus
membrane. Biochem. J., 109: 51-59.

Gurr, M.I. and A.T. James 1971 Lipid biochemistry: an
introduction, Chapman and Hall, London, pp. 125-205.

Haest, C.W.M., J. DeGier and L.L.M. van Deenen 1969 Changes

in the chemical and the barrier properties of the membrane lipids of E. coli by variation of the temperature of growth. Chem. Phys. Lipids, 3: 413-417.

Hooper, S.N. and R.G. Ackman 1971 Trans-6-hexadecenoic acid and the corresponding alcohol in lipids of the sea anemone Metridium dianthus. Lipids, 6: 341-346.

Jefferts, E., R.W. Morales and C. Litchfield 1974 Occurrence of cis-5, cis-9-hexacosadienoic and cis-5,cis-9,cis-19-hexacosatrienoic acids in the marine sponge Microciona prolifera. Lipids, 9: 244-247.

Kates, M. and B.E. Volcani 1966 Lipid components of diatoms. Biochim. Biophys. Acta, 116: 264-278.

Klenk, E., W. Knipprath, D. Eberhagen and H.P. Koof 1963 Über die ungesättigten fettsäuren der fettstoffe von süsswasser- und meeresalgen. Z. Physiol. Chem., 334:44-59.

Korn, E.D. 1964 The fatty acids of Euglena gracilis. Lipid Res., 5: 352-362.

Korn, E.D., C.L. Greenblatt and A.M. Lees 1965 Synthesis of unsaturated fatty acids in the slime mold Physarum polycephalum and the zooflagellates Leishmania tarentolae, Trypanosoma lewisi, and Crithidia sp. : a comparative study. J. Lipid Res., 6: 43-50.

Litchfield, C., A.G. Greenberg, G. Noto and R.W. Morales 1975 Unusually high levels of C_{24}-C_{30} fatty acids in sponges of the class Demospongiae. Unpublished.

Longley, R.P., A.H. Rose and B.A. Knights 1968 Composition of the protoplast membrane from Saccharomyces cerevisiae. Biochem. J., 108: 401-412.

McKillican, M.E. and R.P.A. Sims 1964 The endosperm lipids of three Canadian wheats. J. Am. Oil Chemists' Soc., 41: 340-344, 554-557.

Morales, R.W. and C. Litchfield 1976 Unusual C_{24},C_{25}, C_{26} and C_{27} polyunsaturated fatty acids of the marine sponge Microciona prolifera. Biochim. Biophys. Acta, in press.

Nichols, B.W. and B.J.B. Wood 1968 The occurrence and biosynthesis of gamma-linolenic acid in a blue-green alga, Spirulia platensis. Lipids, 3: 46-50.

Ray, T.K., V.P. Skipski, M. Barclay, E. Enser and F.M. Archibald 1969 Lipid composition of rat liver plasma membranes. J. Biol. Chem., 244: 5528-5536.

Schlenk, H. and J.L. Gellerman 1965 Arachidonic, 5,11,14, 17-eicosatetraenoic and related acids in plants: identification of unsaturated fatty acids. J. Am. Oil Chemists' Soc., 42: 504-511.

Singer, S.J., G.L. Nicholson, and L. Barth 1972 Fluid mosaic model of the structure of cell membranes. Science, 175: 720-731.

Sipos, J.C. and R.G. Ackman 1968 Jellyfish (Cyanea capillata) lipids: fatty acid composition. J. Fish. Res. Bd. Canada, 25: 1561-1569.

Subrahmanyam, D., L.B. Moturu and R.H. Rao 1971 On the phospholipids of Culex pipiens fatigans. Lipids, 6: 867-872.

Ulsamer, A.G., P.L. Wright, M.G. Wetzel and E.D. Korn 1971 Plasma and phagosome membranes of Acanthamoeba castellanii. J. Biol. Chem., 51: 193-215.

Varanasi, U. and D.C. Malins 1974 The structure of phospholipids and triacylglycerols containing long-chain iso acids in the porpoise Tursiops gilli. Biochim. Biophys. Acta, 348: 55-62.

Watanabe, T. and R.G. Ackman 1974 Lipids and fatty acids of the American (Crassostrea virginica) and European flat (Ostrea edulis) oysters from a common habitat, and after one feeding with Dicrateria inornata or Isochrysis galbana. J. Fish. Res. Bd. Canada, 31: 403-409.

Taxonomy and Ecology

A TAXONOMIC STUDY OF THE SPONGILLA ALBA, S. CENOTA,
S. WAGNERI SPECIES GROUP (PORIFERA: SPONGILLIDAE)
WITH ECOLOGICAL OBSERVATIONS OF S. ALBA

Michael A. Poirrier

Department of Biological Sciences
University of New Orleans
Lake Front, New Orleans, Louisiana 70122

ABSTRACT: The taxonomy of the fresh-water sponges, Spongilla alba, S. cenota and S. wagneri was studied and it was concluded that S. wagneri Potts, 1889 is synonymous with S. alba Carter, 1849. Ecomorphic variation in S. lacustris was discussed and its morphology was compared to that of S. alba and S. cenota Penney and Racek, 1968. The ecology of S. alba in Lake Pontchartrain, Louisiana was investigated. It had an annual life cycle; gemmules hatched in the spring, growth and sexual reproduction occurred during the summer, gemmules formed in the fall, and gemmules remained dormant during the winter. Various physicochemical parameters of its habitat were investigated. Spongilla alba appears to be a good water-quality indicator species because it once occurred throughout Lake Pontchartrain but in recent years has been absent from areas affected by storm water discharge from metropolitan New Orleans, Louisiana. The presence of S. alba in oligohaline waters in North America and prior widespread records from similar habitats on other continents indicate that it probably occurs in such habitats in tropical and subtropical environments throughout the world. S. cenota previously known only from Yucatan, Mexico was reported from southern Florida.

Penney and Racek ('68) in a comprehensive revision of a worldwide collection of fresh-water sponges referred to three poorly known species as the Spongilla alba, S. cenota, S. wagneri group. These sponges were reported as differing from other members of the genus Spongilla by having a cushion-like growth form instead of a branching mode of growth. Their microscleres are also different in having longer spines at the middle of their shafts. They are morphologically similar to S. lacustris, and in North America they have often been regarded as variants of this variable species (Jewell, '59; Old, '36; Poirrier, '65, '69; Rioja, '53). Spongilla alba Carter, 1849 is a well-described species which has been reported from India, S. E. Asia, Africa, Australia and South America (Racek, '69). Spongilla cenota Penney and Racek, 1968 is known only from its type locality in Cenote Xtoloc, near Chichen Itza, Yucatan, Mexico. Spongilla wagneri is known from its type locality, a stream 12 to 15 miles east of Lostman's Key in the Florida Everglades; Lake Pontchartrain, Louisiana (Smith, '22); and Mill Creek State Park, Sumter Co. South Carolina (Penney and Racek, '68). Although they were treated as distinct species, Penney and Racek ('68) mentioned that S. cenota and S. wagneri could possibly be of infraspecific importance with geographic variation from S. alba in South America through S. cenota in Yucatan, Mexico to S. wagneri in Florida and South Carolina. The goals of this study were to provide a better understanding of these species and their relation to S. lacustris.

Data for this report are from studies of Louisiana freshwater sponges which began in 1964. These studies have included taxonomy, ecology, distribution (Poirrier, '65, '69) and starch-gel electrophoresis (Arceneaux, '73). My interest in S. alba has been recently stimulated by studies of its ecology in southern Lake Pontchartrain, Louisiana where its distribution appears to have been affected by runoff from outfall canals draining the greater New Orleans area (Poirrier et al., '75). These studies in Lake Pontchartrain led to an examination of this species group.

Fresh-water sponges are usually restricted to a narrow set of physicochemical conditions and could serve as indicators of water quality (Harrison, '74). A better understanding of their systematics and ecology could result in their more widespread use in water quality studies.

MATERIALS AND METHODS

Slides of S. alba Carter, 1849 in the Annandale, Gee and Penney collections; slides of S. wagneri, Penney and Racek, 1968; slides of S. cenota Penney and Racek, 1968 and other slides in the Smithsonian Institution were examined. Specimens of S. wagneri Potts, 1889 from the Potts collection were borrowed from the Philadelphia Academy of Natural Sciences.

I collected specimens of sponges reported as S. wagneri by Penney and Racek ('68) from that locality in South Carolina in January 1975 and specimens of S. alba and S. cenota from the Florida Everglades near the type locality of S. wagneri Potts, 1889 in March 1975. Specimens of S. alba from Louisiana and S. lacustris from Louisiana and other southern states were obtained on numerous collecting trips from 1964 through 1968.

Sponges were collected by wading in shallow water and by diving in deeper water. They were removed from the substrate with a knife, dried and stored in cardboard boxes. Slides of spicules were prepared by digesting tissue in boiling nitric acid, rinsing with water and alcohol, and then mounting dried spicules in balsam (Pennak, '53). Physicochemical measurements of habitat parameters were made according to the procedures of the American Public Health Association ('71).

RESULTS

Systematics

Spongilla alba Carter, 1849, p. 83 - Penney and Racek, 1968, p. 16; Spongilla wagneri Potts, 1889, p. 7; Spongilla wagneri, Smith, 1922, p. 106; Spongilla lacustris, Moore, 1953, p. 42--(Not Spongilla wagneri = Spongilla lacustris, Penney and Racek, 1968, p. 20).

A study of reference specimens of S. alba Carter, 1849 and S. wagneri Potts, 1889 indicated that they are morphologically indistinguishable. Specimens of S. alba in the Smithsonian Institution collections of Annandale, Gee and Penney were compared to specimens of S. wagneri in the Potts collection in the Philadelphia Academy of Natural Sciences. Specimens which I collected from near the type locality of S. wagneri in southern Florida were also examined. Although specimens from different continents were studied no apparent geographic variation was detected.

The description of S. wagneri by Penney and Racek ('68) was found to be based upon S. lacustris. Their description differs from Potts' original description in the length of the microscleres and gemmoscleres and in many other characteristics including the arrangement of the spines on the gemmoscleres. There is no overlap in the size of microscleres and gemmoscleres given in Penney and Racek ('68) and measurements given in Potts (1889), but spicule dimensions of S. wagneri Potts are very close to those reported for S. alba by Penney and Racek ('68) (Table 1). Moreover sponges were obtained from a fresh-water site, whereas, prior records of S. wagneri were from brackish water.

Specimens of S. alba were obtained from a considerable range of salinity in Lake Pontchartrain, but little ecomorphic variation in spicule morphology was present. There was some variation in the arrangement of spines upon the gemmo-

205

scleres. In low salinity areas there were more gemmoscleres with spines along the shaft in a given gemmule than in high salinity areas where spines were restricted to the tips.

TABLE I

SPICULE DIMENSIONS OF S. ALBA AND
S. WAGNERI IN MICRONS

		Micro-sclere	Gemmo-sclere	Mega-sclere
S. wagneri,	1.	49-62	48-75	144-270
Penney & Racek, 1968	w.	2-4	6-8	7-12
S. wagneri,	1.	68-129	93-151	297-403
Potts, 1887	w.	2-3	6-10	15-22
S. alba,	1.	75-124	75-130	256-420
Penney & Racek, 1968	w.	2-3	5-10	12-22

There was considerable variation in colony morphology. Specimens from low salinity water near the mouth of the Tchefuncte River were typical of those described by Potts (1889) as S. wagneri. These colonies were white, encrusting, two to four mm thick, with numerous microscleres on their surface. The skeletal structure was less compact than larger colonies found in other areas of the lake and the skeleton below the layer of microscleres was loose and open as described by Potts (1889). In most areas of Lake Pontchartrain colonies were two to three cm thick. Most colonies had a cushion-like growth form, but some specimens had finger-like projections (Fig. 1). Developing colonies grew together and often the size of the sponge was limited only by the size of the substrate. It was common for pilings and seawalls to be covered by sponges. Color ranged from white to grey in thin specimens from low salinity areas and from light tan to grey in thick specimens from high salinity areas. Color was influenced by the amount of suspended material in the water. Sponges kept in aquaria changed from grey or tan to white as materials were egested. The number of microscleres varied. They were extremely abundant in colonies from the northern shore near the mouths of rivers and less abundant in higher salinity areas.

Fig. 1. Branching Growth Form of S. alba.

Although S. alba, S. cenota and some ecomorphs of S. lacustris have similar microscleres and megascleres, they can
be distinguished from each other by their characteristic
gemmoscleres. Spongilla alba can be distinguished from S.
lacustris and S. cenota by gemmoscleres which have spines
concentrated at the ends of the shafts (Fig. 2). Spongilla
lacustris has gemmoscleres which are spined uniformly along
the shaft. Considerable variation occurs in their size and
shape as described and illustrated by Potts (1887). Spongilla cenota can be distinguished from S. alba and S. lacustris by its stout gemmoscleres (Fig. 3) with large spines on
both ends. The gemmule pneumatic layer of S. cenota is very
different from its congeners. It has conical projections
covering the surface except near the foraminal aperture.
Around the aperture, there is a definite crater-like de-
pression which is very distinct in gemmules which have been
soaked in water.

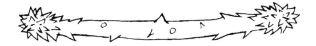

Fig. 2. Gemmosclere of S. alba.

Spongilla cenota, previously known only from Yucatan, Mexico occurs in southern Florida. Specimens in the Smithsonian
Institution collected in 1901 by J. E. Benedict from the
Miami River seven mi from its mouth and identified as S.
wagneri by N. Annandale are S. cenota. I collected it from

the Tamiami Canal off U.S. Hwy 41, 2 mi east of Fla. Hwy 94 in the Everglades on March 28, 1975. It was found on limestone rock associated with Ephydatia fluviatilis. Spicular measurements of S. cenota from Florida were comparable to values reported by Penney and Racek ('68) for the Mexican population.

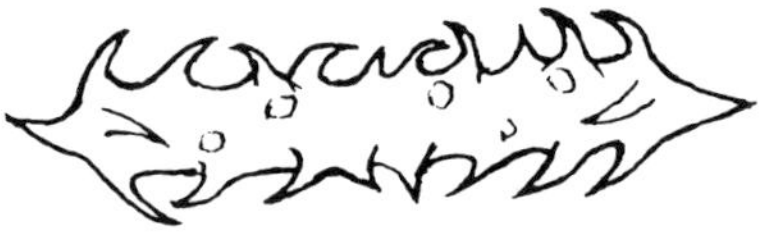

Fig. 3.   Gemmosclere of S. cenota.

ECOLOGY

In a study of Louisiana Spongillidae (Poirrier, '69) twelve species were obtained from 184 localities throughout Louisiana. Spongilla alba was only found in oligohaline estuaries in the coastal marshes. These included Lake Pontchartrain where collections were obtained from near-shore sites in all areas of the lake. While general observations began as early as 1960 detailed studies were conducted during 1964 (Poirrier, '65) and additional data were obtained in the summer of 1973 (Poirrier et al., '75).

Because S. alba occurred throughout Lake Pontchartrain and was absent from adjacent estuaries to the east and west, a characterization of physicochemical factors associated with the occurrence of S. alba would correspond to the physicochemical characteristics of Lake Pontchartrain. Darnell ('58) gives a description of Lake Pontchartrain from studies conducted during 1953 through 1955. It has an oval shaped basin with a surface area of 635 sq mi and an average depth of 11 ft. The average salinity was less than 6 ppt and varied with rainfall and the influx of more saline waters from the Gulf of Mexico. He reported the temperature as ranging from $9^{\circ}C$ in December to $34^{\circ}C$ in June and August. Shallow water temperatures were higher ranging up to $39^{\circ}C$. The most rapid rise in temperature occurred in April or May, and the most rapid decrease occurred in early November. Because Lake Pontchartrain is shallow it can become turbid from heavy wave action. During winter storms, Secchi disc readings of less than one inch were recorded. Overall Secchi disc visibility averaged about 4 ft. with average summer visibilities ranging about 6 ft. and winter visibilities ranging about 2 to 3 ft.

Gemmules hatched in the spring, during May, as water temperatures increased from 22 to 25°C. Sponges first spread out as a thin layer and then rapidly grew into colonies 2 to 3 cm thick by late June. Specimens collected during July and August and placed in aquaria were observed releasing larvae. At this time sponges also appeared upon substrate previously unoccupied by sponges. They grew throughout the summer and spread to almost cover every available substrate. Structures such as pilings and seawalls became completely covered with sponges. Gemmulation occurred during September and October. Soon after, colonies disintegrated and all that was left during the winter months was a layer of gemmules.

Physicochemical data were obtained during the summers of 1964 and 1973. Data obtained in 1973 included pH values and alkalinities which are higher than those normally found in Lake Pontchartrain. These high values resulted from alkaline Mississippi River water entering Lake Pontchartrain during the 1973 opening of the Bonnet Carré Spillway. Spongilla alba was found living within the following range of physico-chemical conditions:

pH	6.3 - 9.0
Total Alkalinity	18 - 78 ppm
Phenolphthalein Alkalinity	0 - 21 ppm
Chloride	0.036 - 3.20 ppt
Specific Conductance	230 - 7407 umhos/cm^2
Temperature	25 - 37°C

In Lake Pontchartrain S. alba is part of an estuarine epifaunal community. It was found associated with marine organisms such as: Bimeria fransciscana, Victorella pavida, Membranipora sp., Neanthes succinea, Polydora sp., Balanus improvisus, Corophium lacustre, Rhithropanopeus harrisii, Congeria leucophaeta, Crassostrea virginica and Brachidontes recurvus. The fresh-water sponge Trochospongilla leidyi was found associated with S. alba in low salinity areas.

DISCUSSION

Systematics

The apparent synonomy of S. wagneri with S. alba was obscured by inaccurate descriptions of workers since Potts (1889). Although S. wagneri was only reported from Florida, Louisiana, South Carolina and possibly from Rio de Janeiro (Weltner, 1885; Gee, '32) descriptions and discussions of past workers are confusing. Jewell ('59) did not include S. wagneri in her key to North American fresh-water sponges and apparently considered it a variety of S. lacustris. Pennak

('53) included S. wagneri in his key but separated it from other species by smooth microscleres, a character which has never been reported in this species. Eshleman ('50) only reported it from the type locality but contrary to Potts' description stated that the skeletal spicules are sometimes microspined. Penney and Racek ('68) added to this confusion by describing what is probably a variety of S. lacustris as S. wagneri.

Specimens of these sponges in the Penney collection and sponges which I obtained from the same locality in South Carolina are similar to forms of S. lacustris from Louisiana and other southern states. They never have the characteristic branching of northern forms and have considerable variation in the size and morphology of microscleres, the thickness of the pneumatic layer and the size and structure of the gemmoscleres. In soft, acid waters the gemmoscleres are small and the pneumatic layer is either absent or slightly developed. The gemmoscleres are absent or vary in size and number with the development of the pneumatic layer. These variations are ecomorphic and depend on the chemical properties of the habitat water. A detailed discussion of field and laboratory studies of ecomorphic variation in S. lacustris will be presented in a subsequent publication. The sponges described by Penney and Racek ('68) appear to be southern forms of S. lacustris. There may be genetic differences between northern and southern populations of S. lacustris due to evolutionary processes, but additional studies are needed to document this.

Spongilla alba cannot be regarded as a brackish-water ecomorph of S. lacustris. Besides the morphological differences already presented, additional differences between S. lacustris and S. alba are apparent in the discussion of their ecology to follow and in protein electrophoresis studies of Louisiana Spongillidae (Arceneaux, '73). Spongilla alba was always distinguished from S. lacustris by differences in esterase, leucine aminopeptidase and glutamate oxalotransaminase.

Specimens of S. alba from North America were very similar to specimens in the Smithsonian Institution collected from other areas of the world. As reported by Racek ('69) there does not appear to be any significant spicular difference in specimens from different continents.

The synonymy of S. wagneri with S. alba and the presence of S. alba and S. cenota in southern Florida does not support the hypothesis of Penney and Racek ('68) that infraspecific structural changes from S. alba in South America through S. cenota in Yucatan, Mexico to S. wagneri in Florida and South Carolina may exist. A comparison of specimens of S. cenota from three localities supports its designation as a distinct species (Penney and Racek, '68). Specimens from the three sites were quite similar in morphology and morphometry indicating that it is not an ecomorph or subspecies of S. alba or S. lacustris.

ECOLOGY

In Lake Pontchartrain, Louisiana S. alba has an annual life cycle with gemmules hatching in the spring, rapid growth and sexual reproduction during the summer and gemmule formation in the fall. Gemmule hatching appears to be related to increasing temperature during the spring. Gemmule formation appears to be related to decreasing temperature and increasing turbidity which occur in Lake Pontchartrain during the fall.

Even though S. alba was collected from marsh canals and tidal streams which flow into Lake Pontchartrain, it was never found in true fresh water. The lower range of specific conductance and chloride values reported in this study were from the tidal areas of the Tchefuncte River and are probably lower than mean conditions at that site. The maximum salinity tolerance of S. alba may be higher than indicated by these values because its occurrence in neighboring, higher-salinity estuaries appears to be limited by turbidity. It can undoubtedly withstand higher salinities for short periods because salinities as high as 18.6 ppt have been recorded in Lake Pontchartrain after storms (Darnell, '58).

Spongilla alba appears to be restricted to low salinity waters and probably has a worldwide distribution in such habitats. It appears to be more common in tropical and subtropical climates but may also occur in temperate latitudes. Records of S. lacustris from brackish water (Tendal, '67; Gosner, '71) should be re-examined because they may be based on S. alba.

Ecological differences between S. alba and S. lacustris provide additional evidence for regarding S. alba as a distinct biological species and not a brackish-water ecomorph of S. lacustris. In Louisiana, S. lacustris occurred in acid, fresh-water habitats low in dissolved solids. These habitats included streams, ditches and ponds. Although S. lacustris occurred in the upper reaches of streams that flow into Lake Pontchartrain and S. alba occurred in the tidal areas, neither occurred in midcourse areas. Attempts to introduce S. alba to fresh water and S. lacustris to brackish water under both field and laboratory conditions were unsuccessful. Spongilla lacustris lives throughout the year, while S. alba is only active during the warm months.

When S. alba was studied in 1964, it was extremely abundant on the New Orleans Seawall and the Lake Pontchartrain Causeway. Since that time it has decreased in abundance. It no longer occurs along the southern end of the causeway and has an erratic distribution along the seawall. These changes in distribution appear to be related to recent changes in the water quality of the southern sector of the lake which have probably been brought about by discharge from outfall canals in the greater New Orleans area (Poirrier et al., '75).

ACKNOWLEDGMENTS

This work was supported by a federal grant under P. L. 88-379, the Water Resources Research Act of 1964, which was made to the Louisiana Water Resources Research Institute by the Office of Water Resources Research of the U. S. Department of the Interior (Project No. A-035-LA). I acknowledge the assistance of S. Fuller, Philadelphia Academy of Natural Sciences, and K. Ruetzler, Smithsonian Institution, in obtaining loans of museum specimens.

LITERATURE CITED

American Public Health Association 1971 Standard Methods for the Examination of Water and Wastewater. 13th ed., American Public Health Assoc., New York.

Arceneaux, Y. A. 1973 Application of starch-gel electrophoresis to the taxonomy of the Spongillidae: Porifera. M. S. Thesis, University of New Orleans, New Orleans, Louisiana, 45 p.

Carter, H. J. 1849 A descriptive account of the fresh-water sponges (Genus Spongilla) in the Island of Bombay, with observations on their structure and development. Ann. Mag. Nat. Hist., 4:81-100.

Darnell, R. M. 1958 Food habits of fishes and larger invertebrates of Lake Pontchartrain, Louisiana, an estuarine community. Publ. Inst. Marine Sci., Univ. Texas, 5: 353-416.

Eshleman, S. K. 1950 A key to Florida's fresh-water sponges, with descriptive notes. Quart. Journ. Fla. Acad. Sci., 12:36-44.

Gee, N. G. 1932 The known fresh-water sponges. Peking Nat. Hist. Bull., 6:25-51.

Gosner, K. L. 1971 Guide to the Identification of Marine and Estuarine Invertebrates. Wiley-Interscience, New York.

Harrison, F. W. 1974 Sponges (Porifera: Spongillidae).
In: Pollution Ecology of Freshwater Invertebrates. C. W.
Hart and S. Fuller, eds., Acad. Press, New York, pp. 29-66.
Jewell, M. E. 1959 Porifera. In: Ward and Whipple, Fresh-
water Biology. (W. T. Edmondson, ed.) 2nd ed., John Wiley,
New York, p. 298-312.
Moore, W. G. 1953 Notes on Louisiana Spongillidae. Proc.
La. Acad. Sci., 16:42-43.
Old, M. C. 1936 Yucatan Fresh-water Sponges. In: Pearse,
Creaser and Hall, The Cenotes of Yucatan. Carnegie Inst.
Wash. Publ. No. 457:29-31.
Pennak, R. W. 1953 Fresh-water Invertebrates of the United
States. Ronald Press Co., New York.
Penney, J. T. and A. A. Racek 1968 Comprehensive revision
of a world-wide collection of freshwater sponges (Pori-
fera: Spongillidae). U. S. Nat. Mus. Bull., 272:1-184.
Poirrier, M. A. 1965 Ecological variation in Spongilla la-
custris (L.). M. S. Thesis. Louisiana State University,
Baton Rouge, Louisiana, 58 p.
_____1969 Louisiana fresh-water sponges: ecology, taxono-
my, and distribution. Ph. D. Dissertation, Louisiana
State University, Univ. Microfilms Inc., Ann Arbor, Michigan, No. 70-9083.
Poirrier, M. A., J. S. Rogers, M. A. Mulino, and E. S. Eisen-
berg 1975 Epifaunal invertebrates as indicators of water
quality in southern Lake Pontchartrain. La. Water Resources Research Inst., Technical Report No. 5., Louisiana
State University, Baton Rouge, Louisiana.
Potts, E. 1887 Contributions toward a synopsis of the Amer-
ican forms of freshwater sponges. Proc. Acad. Nat., Sci.,
Phila., 39:158-279.
_____1889 Report upon some freshwater sponges collected in
Florida by Jos. Willcox, Esq. Trans. Wagner Inst., 2:5-7.
Racek, A. A. 1969 The fresh-water sponges of Australia
(Porifera: Spongillidae). Aust. J. Mar. Freshwater Res.,
20:267-310.
Rioja, E. 1953 Estudios hidrobiologicos. XI. Contribucion
al estudio de las esponjas de agua dulce de Mexico. An.
Inst. Biol. Mex., 24:425-433.
Smith, F. 1922 A new locality for Spongilla wagneri Potts.
Trans. Amer. Micros. Soc., 41:106.
Tendal, O. S. 1967 On the fresh-water sponges of Denmark.
Videnskabelige Meddeleser Fra Dansk Naturbistorisk For-
ening., 130:173-178.
Weltner, W. 1895 Spongillidenstudien. III. katalog and
verbreitung der beksannten süsswasserschwämme. Arch. f.
Naturg., 61:114-144.

A CYTOLOGICAL STUDY OF THE HALICLONIDAE AND THE CALLYSPONGIIDAE (PORIFERA, DEMOSPONGIAE, HAPLOSCLERIDA)

Shirley A. Pomponi

University of Miami
Rosenstiel School of Marine and Atmospheric Science
4600 Rickenbacker Causeway
Miami, Florida   33149

ABSTRACT: A cytological study of the sponge families Haliclonidae and Callyspongiidae was made to provide a non-skeletal characteristic to delineate taxonomic categories further and to produce a better classification. Specimens of Haliclona viridis, H. rubens, H. variabilis, Dasychalina cyathina, Callyspongia plicifera, and C. vaginalis were collected and prepared for cytological study. Sections were stained to reveal carbohydrates (including glycogen), acid mucopolysaccharides, and nucleic acids. Choanocytes in every species contain an acid mucopolysaccharide, probably mucin which is secreted and distributed throughout the sponge. It functions as a protective covering and/or as a mucous net to trap food particles. Glycogen occurs in most cells and may serve in the storage of reserves. In choanocytes it may function as an energy source for the beating flagella. "Nucleolate" cells occur in all species studied and are multi-functional (ingestion, digestion, storage, elaboration of ground substance). Fiber cells occur in all species. Their location and arrangement is diagnostic. Special cell types, granular cells, are also diagnostic. Granular cells occur among spongin fibers in Haliclona variabilis, Dasychalina cyathina, Callyspongia plicifera, and C. vaginalis, and may function in spongin fibrillogenesis. These cytological findings support biochemical and morphological observations and indicate that the Haliclonidae and Callyspongiidae should remain as two distinct families, but that Haliclona variabilis and Dasychalina cyathina are intermediate species between the two families.

Delimitation of taxonomic categories within the Porifera is extremely difficult due to the variability of characters employed and to the intergradations of these characters among species, genera, families, and orders. Moreover, variability occurs not only among different species but also among individuals of the same species. While species distinctions

215

based on size, shape, consistency, color, and skeletal architecture are useful in field identifications, these characters are often so variable that their taxonomic value is questionable. In fact, the weight of a particular character may vary from one taxon to another. For example, external shape may be of little importance in one species (e.g., Haliclona viridis) and of diagnostic value in another (e.g., Callyspongia plicifera).

Attempts have been made during the past forty years to produce a better classification based on cytological, biochemical (Bergmann, '49, '62; Bergquist and Hartman, '69; Bergquist and Hogg, '69), and embryological (Lévi, '56, '57) research. Many studies have concentrated on one or two orders (viz., Poecilosclerida and Dendroceratida).

The Haliclonidae and Callyspongiidae, two families within the order Haplosclerida, were chosen for the present cytological study because their simplicity of architecture and spicule content presents systematic problems. The callyspongiids have been grouped with the Haliclonidae by Wiedenmayer (in press) who considered skeletal differences unimportant. This family was included in my study to ascertain if Wiedenmayer's grouping is cytologically valid.

Cytological research has been conducted on other orders, notably Poecilosclerida (Wilson and Penney, '30; de Laubenfels, '32; Simpson, '63, '68a; Borojević and Lévi, '64), Dendroceratida (Lévi, '56; Vacelet, '67; Tuzet et al., '70), Dictyoceratida (Tuzet and Pavans de Ceccatty, '58; Thiney, '72), Tetractinellida (Pavans de Ceccatty, '57; Pavans de Ceccatty and Garrone, '71; Garrone, '71), Hadromerida (Fauré-Fremiet, '31; Cheng et al., '68a, '68b), Epipolasida (Tuzet and Paris, '57), and Choristida (Tuzet and Pavans de Ceccatty, '53). In contrast, little is known of the cytology of the Haplosclerida (Tuzet, '32; Tessenow, '68; Fell, '69, '70; Garrone, '69a, '69b; Schmidt, '70; Harrison, '72a, '72b, '74; Harrison and Cowden, '75; Harrison et al., '74).

My main objectives in studying the histology and cytology of the Haliclonidae and Callyspongiidae were to identify cell types and determine possible functions, and to use this information to examine the taxonomy of sponges in the light of present relationships.

MATERIALS AND METHODS

The species chosen for study were Haliclona viridis, H. rubens, H. variabilis, Dasychalina cyathina, Callyspongia plicifera, and C. vaginalis. All specimens were collected either from the sublittoral zones off Big Pine Key and Key

Biscayne, Florida, at depths of three to five feet, or from reefs off Freeport, Grand Bahama Island, at depths of forty to fifty feet. Specimens were fixed in ethanol-acetic acid (3:1) or Bouin's fluid, embedded in paraffin, sectioned at 8 μm, and stained for cytological study. Tissues fixed in ethanol-acetic acid (3:1) were stained by the following methods: azure B bromide (Flax and Himes, '52) – ribonuclease (Luna, '68) for localization and differentiation of DNA and RNA; Feulgen (Pearse, '68) for determination of DNA; alcian blue (Pearse, '68) for localization of acid mucopolysaccharides; Best's carmine stain for glycogen (Pearse, '68); and Taft's method (Luna, '68) for localization and differentiation of nucleic acids.

Tissues fixed in Bouin's fluid were stained with hematoxylin and eosin to show general organization and certain cytological features, and with PAS-alcian blue (Pearse, '68) for differentiation of carbohydrates from acid mucopolysaccharides.

Spicule preparations were made by boiling a small portion of the surface and mesenchyme of each specimen in nitric acid (Hechtel, '65).

Cell types were defined and identified by their size, shape, nuclear material, and type and number of inclusions. The functions of each cell were predicted by the staining affinities of the cytoplasm and the nature of the cytoplasmic inclusions.

RESULTS

Family Haliclonidae de Laubenfels

The members of the family Haliclonidae typify the order Haplosclerida. The megascleres are simple--diacts of the same type throughout the sponge. There are no microscleres. The spicules are arranged in spongin, and both the proportion of spicules to spongin and the arrangement of spicules in spongin are diagnostic on the generic and specific level. The fibers have a reticulate structure--branching and anastomosing, forming a two- or three-dimensional net. The absence of surface specialization (finer, secondary reticulation of fibers between the meshes of wider fibers at the surface) further distinguishes this family. Wiedenmayer (in press) does not consider the presence or absence of surface specialization diagnostic at the family level, and he drops the family Callyspongiidae including them with the haliclonids. While some species of haliclonids have the tendency to produce a finer reticulation of fibers at the surface,

this is generally not as pronounced as the secondary reticulation of fibers characteristic of callyspongiids. For this
reason, as well as the cytological complements of the
species studied, I have chosen to retain the two families.
However, the reader is referred to Wiedenmayer's excellent
discussion of this problem.

Haliclona viridis (Duchassaing and Michelotti,
1864, p. 81) de Laubenfels, 1936, p. 42.

Morphology: This species is thickly encrusting to ramose
in shape. It is most often dark green, but may also be
bluish-green, brownish-green, or grayish-green. When
handled, the sponge releases a green exudate. The sponge is
soft, smooth, and friable. The surface is punctiform, with
the ostia uniformly distributed about 1 mm apart and ranging
in diameter from 20 to 40 μm. The oscules are about 2 cm
apart and range from 0.3 to 10 mm in diameter. Occasionally
the rims of the oscules are raised 1 mm.
As is characteristic of the Haliclonidae, there is no surface specialization. In some specimens, the fibers protrude
from the surface, making it finely hispid. The reticulation
is isotropic (a disoriented arrangement of spicules and
fibers). The fibers are arranged into irregular bundles.
The spicules are hastate oxeas (Fig. 1) which are straight
to slightly curved and measure 126.7-144.5-159.7 X 2.9-4.5-
6.5 μm (means and extremes of 20 spicules). The oxeas occur
throughout the mesenchyme and protrude through the fibers.

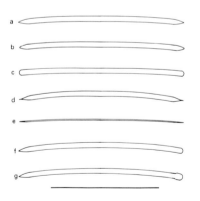

Fig. 1. Spiclues. a-hastate
oxea. b-strongyloxea. c-
strongyle. d-mammiform
oxea. e-raphide. f-style.
g. tylostyle. Scale 100
μm.

Cytology: The cells occurring in the adult of Haliclona
viridis include"nucleolate"cells, choanocytes, and fiber
cells. With the exception of fiber cells, these cells are
not confined to one region.

"Nucleolate" cells are the classical archeocytes, with vescicular nucleus, well-defined pseudopodia, and a variable number and kind of cytoplasmic inclusions. These inclusions generally contain acid mucopolysaccharides. Histochemical characteristics suggest that both the nucleus and cytoplasmic granules contain DNA and RNA. Some granules are glycogen-positive. Smaller, irregular granules are basophilic and eosinophilic. There are also vacuoles which may be clear or may contain an inclusion. The size of the "nucleolate" cell is 5.9-8.0-14.7 X 4.4-5.9-7.3μm; the nucleus is 2.9-3.5-4.5μm.

Choanocytes are scattered throughout the mesenchyme. They are either arranged in chambers (Fig. 2) or occur singly. The cells have a basal, spherical, anucleolate nucleus. Most choanocytes contain a single, large, cytoplasmic inclusion which is strongly acid mucopolysaccharide-positive. Several choanocytes contain, in addition, a glycogen inclusion, and nucleic acids can be localized.

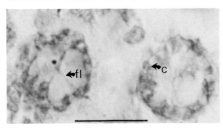

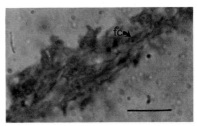

Fig. 2. Choanocyte chambers. c-choanocyte, fl-flagellum. Carnoy, alcian blue. Scale 20 μm.

Fig. 3. Fiber cell bundle characteristic of Haliclona viridis. Fiber cells (fc) lying parallel to each other. Bouin's, hematoxylin and eosin. Scale 20 μm.

Fiber cells are elongate; they lie parallel to each other to form short fiber cell bundles (Fig. 3). My observations suggest that these cellular bundles are characteristic of Haliclona viridis. The fiber cells may contain nucleoli, but usually do not. Rather, the nucleus has irregular chromatin granules. Both DNA and RNA are present in the nucleus. Inclusions are lacking in most cells.

Spongin fibers are scarce, which accounts for the fragility of this species. Spicules are scattered among the cells and ground substance of the mesenchyme.

Pinacocytes are not evident. There is a thin protoplasmic layer with occasional nuclear bulges. Cell membranes are difficult to discern in the very attenuated cytoplasm.

Haliclona rubens (Pallas, 1766, p. 389) de Laubenfels,
    1932, p. 59.
= Haliclona (Amphimedon) compressa (Duchassaing and
    Michelotti) Wiedenmayer, 1974, p. 367*.

Morphology: Haliclona rubens is a common West Indian
sponge and easily identifiable. It is a deep carmine red
and retains this color in alcohol. It branches from an
amorphous base. The branches range in size from 8 to 40 cm
in height and 2 to 4 cm in diameter. The tips of the branches are pilose due to protruding spongin fibers. The consistency is firm and tough but elastic. The surface of the
sponge is smooth. Oscules are irregularly distributed and
measure from 3 to 8 mm in diameter. There is no surface
specialization.
    The spicules are hastate oxeas, strongyloxeas, and a few
stongyles (Fig. 1), with a size range of 113.0-128.0-138.0
X 2.6-3.9-5.0 μm. These spicules are arranged in plurispicular tracts of reticulating spongin fibers which range
in width from 25 to 300 μm.
    Cytology: "Nucleolate" cells, choanocytes, and fiber cells
are found in Haliclona rubens. "Nucleolate" cells are variable in size (5.9-8.6-14.7 X 2.9-6.0-8.8 μm), type of inclusions, and staining affinities. Many are DNA- and/or RNA-
positive. Some inclusions contain acid mucopolysaccharides
or carbohydrates, and some granules are eosinophilic and
basophilic. Clear vacuoles are often found within the "nucleolate" cells.
    Choanocytes occur throughout the sponge except near the
surface. They are usually clustered in chambers and not
scattered. The nucleus is spherical,basal, and anucleolate,
but often granular, and it stains for nucleic acids with
azure B bromide. The cytoplasmic inclusions contain acid
mucopolysaccharides and carbohydrates
    Fiber cells are located in the reticular fiber bundles
suggestive of loose connective tissue. They are elongate
and the cytoplasmic extensions are PAS-positive. The nucleus
contains chromatin granules and is DNA- and RNA-positive.
    Both spongin A and B fibers are abundant. The coarse
spongin B is scattered throughout the sponge and the finer
spongin A fibers are located in the reticular fiber bundles.

---

*Wiedenmayer (in press) presents an extensive revision of
    sponges of the western Bahamas. In a recent paper (Wieden-
    mayer, '74) he lists, without synonymies, replacement names
    for three of the species included in the present study.

<u>Haliclona variabilis</u> (Dendy, 1890, p. 353) de Laubenfels,
1950, p. 42.

<u>Morphology</u>: <u>Haliclona variabilis</u> is a difficult sponge to
identify with any certainty. This species is often confused
with <u>H</u>. <u>subtriangularis</u> (Duchassaing and Michelotti) de
Laubenfels, 1932. The specimens I identified correspond
with the description of <u>Haliclona variabilis</u> by de Laubenfels
('50). This species is flabellate or ramose with irregular
branches. The color varies from beige to light blue-green
or lavender in a single specimen. The consistency is tough
and stiff.

The surface of the sponge is slightly hispid to conulose.
Ostia are irregularly scattered with diameters ranging from
200 to 500 μm. Oscules are also randomly distributed and
measure from 1 to 5 mm in diameter. The tips of the branches
are pilose.

The spongin fibers are arranged in more definite tracts
than those of <u>H</u>. <u>viridis</u> or <u>H</u>. <u>rubens</u>. They form a more or
less patterned reticulation and appear to ascend from a base
point. In some areas, there is a finer reticulation of
fibers on the surface. Isodictyal reticulation is characteristic, with the spicules joined at the tips by spongin,
forming triangular meshes. The spicules are oxeas and raphides (Fig. 1). Most of the oxeas are hastate but some have
mammiform ends. Both the oxeas and raphides are straight or
slightly curved. The sizes are 183.8-<u>203.1</u>-226.3 X 4.0-<u>5.7</u>-
8.5 μm for oxeas and 155.0-<u>168.7</u>-187.5 X 1.0 μm for raphides.

<u>Cytology</u>: "Nucleolate" cells, choanocytes, pinacocytes,
fiber cells, and a special cell type--granular cells--occur-
red in the specimens of <u>Haliclona variabilis</u> examined. The
"nucleolate" cells are similar to those described for <u>H</u>.
<u>viridis</u>, with a size range of 5.9-<u>7.6</u>-11.8 X 4.4-<u>6.1</u>-8.8 μm.
The cytoplasm contains acid mucopolysaccharide inclusions,
and smaller, irregular, basophilic and eosinophilic granules.
The nucleus, nucleolus, and cytoplasmic granules and inclusions contain nucleic acids. In some cells, there is a vacuole which may have an acid mucopolysaccharide granule within.

Choanocytes occur singly and in chambers throughout the
sponge, except near the surface. They contain an acid mucopolysaccharide inclusion, and in some cells there is also a
clear vacuole. The nucleus is spherical, lacks a nucleolus
and is located at the base of the cell.Nucleic acids are
present in the nucleus and cytoplasm.

Cells which may be pinacocytes are flattened, and are distinguishable in only a few areas of the surface. It could
not readily be determined whether the cells are individual

or form a syncytium. The cytoplasm is basophilic, and the nucleus, when visible lacks a nucleolus.

Fiber cells occur in bundles, but the bundles are more fibrous than cellular, which seems to be a characteristic of the tougher, more rigid species. The nucleus may contain a nucleolus. Basophilic granules give the cytoplasm a beady appearance. Both the nucleus and granules contain nucleic acids.

There are two types of spongin fibers--coarse spongin B fibers and the more delicate spongin A. The latter forms the reticular fiber bundles. These bundles resemble loose connective tissue of vertebrates. Spicules are enclosed within the spongin B fibers.

The special cells, granular cells (Fig. 4), contain several large inclusions from 1 to 2 μm in diameter. These inclusions fill the cell and are strongly PAS- and glycogen-positive. The cells are spherical and pseudopodia have not been observed. The dimensions are 5.9-9.8-16.2 X 4.4-6.5-10.1 μm. The granular cells occur in the retucular spongin A fibers (Fig. 5).

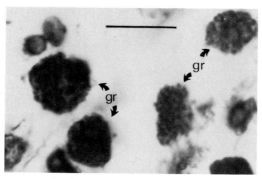

Fig. 4. Granular (gr) cells filled with glycogen inclusions. Bouin's, PAS-alcian blue. Scale 20 μm.

Dasychalina cyathina de Laubenfels, 1936, p. 45
= Niphates digitalis (Lamarck) Wiedenmayer, 1974, p. 367.

Morphology: Dasychalina cyathina is a hollow tube, often compressed. The atrium is often open at both the top and bottom of the sponge. The diameter of the top is 1.5 to 5 times greater than the diameter at the base. The atrium is fringed with fibrous projections, 5 to 15 mm long. Occasionally this species may be solid and globular or ramose. The texture is stiff and tough. The color is lavender or light blue. The outer surface is microscopically hispid and often

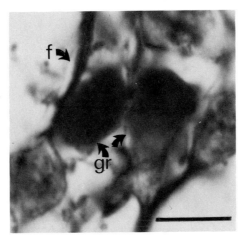

Fig. 5. Granular cells (gr) among reticular spongin A fibers
(f). Bouin's, PAS-alcian blue. Scale 20 μm.

beset with specimens of the anemone Parazoanthus. The inner
surface is smooth, with oscules 0.4 to 1.5 mm in diameter.

The fibers form coarse reticulations and give a honeycombed
appearance externally. They ascend at the fringe surrounding
the atrium. Their width varies from 20 to 80 μm. The spicules are oxeas, measuring 158.1-189.9-219.4 X 2.9-5.8-8.8 μm,
and styles (Fig. 1) 147.5-170.5-195.6 X 4.0-5.5-7.1 μm, some
with tylostylote modifications. The reticulation is isotropic. The spicules are arranged in plurispicular tracts
and are scattered throughout the mesenchyme. Styles have
not been reported in previous species descriptions, but the
numbers of spicules present and the fact that they occur in
every specimen collected indicate that they are not foreign
spicules.

Cytology: This sponge has "nucleolate" cells, choanocytes,
fiber cells, and the special cells, granular cells. The
"nucleolate" cells (Fig. 6) tend to be elongated (5.9-8.1-
11.8 X 4.4-6.3-8.8 μm). The nucleus contains DNA and the
nucleolus RNA. Larger cytoplasmic inclusions contain acid
mucopolysaccharides and carbohydrates, including glycogen,
while smaller inclusions are basophilic. Some cells contain
vacuoles.

The choanocytes are grouped in chambers or occur singly.
The basal nucleus is spherical, and chromatin granules are
sometimes evident. Cytoplasmic inclusions contain acid
mucopolysaccharides and carbohydrates, including glycogen.
Both the nucleus and the cytoplasmic granules contain nucleic
acids.

223

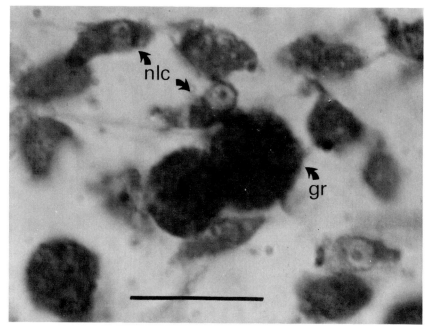

Fig. 6. "Nucleolate" cells (nlc) and granular cells (gr).
Bouin's, PAS-alcian blue. Scale 20 μm.

The fiber cells are elongate and are found in reticular
bundles resembling loose connective tissue. The nucleus is
spherical and contains chromatin granules which are DNA- and
RNA-positive. Cytoplasmic extensions contain carbohydrates,
which may or may not be glycogen.

The special cell type, or granular cell (Fig. 6), is characterized by PAS-positive and glycogen-positive inclusions
which fill the cell and occasionally mask the nucleus. The
cell dimensions are 5.9-10.1 17.6 X 4.4-7.8-8.8 μm. The
cells occur near spongin B fibers and among the fiber cells
in the reticular bundles.

Both spongin A and spongin B fibers are filled with spicules, forming definite plurispicular tracts.

Family Callyspongiidae de Laubenfels

The callyspongiids differ from the haliclonids in that they
have a secondary surface reticulation (Fig. 7). Between the
wide meshes of spongin fibers at the surface of the sponge,
there occurs a finer reticulation of spongin fibers. Di-
actinal megascleres are present, although they are generally

smaller and less abundant than in the haliclonids. Spongin
B fibers are more abundant than in the Haliclonidae.

Fig. 7. Secondary surface
reticulation. Finer reticulation of thin spongin fibers
between the larger meshes formed by the wider spongin fibers.
Characteristic of the Callyspongiidae.

Callyspongia plicifera (Lamarck, 1814, p. 436) de
Laubenfels, 1950, p. 61.

Morphology: This sponge is a cylinder, growing singly or
forming coalescent tubes , and is usually hollow but may be
solid. The color is iridescent pink, orange, blue, or
purple. It is very stiff and tough. The outer surface consists of distinct ridges (from which the species takes its
name). The inner surface is smooth and has oscules irregularly scattered. The atrium is generally narrow at the top,
forming a rim.
The fibers are arranged in layer upon layer of reticulations. The wider fibers form the ridges. Interiorly, the
fibers form longitudinal tracts. Spicules are very scarce
and, when present, are small oxeas, 79.4-102.2-139.4 X 2.1-
2.9-3.6 µm.
Cytology: This sponge contains "nucleolate" cells, choanocytes, fiber cells, and the special cells, granular cells.
The "nucleolate" cells (Fig. 8) range in size and shape from
small, spherical cells to larger, elongate cells (5.9-9.0-
13.2 X 4.4-6.5-7.3 µm). The nucleus is stained by methods
for demonstration of DNA and the nucleolus is stained for
RNA. The cytoplasm is granular, basophilic, and DNA- and
RNA-positive. Larger inclusions contain acid mucopolysaccharides.
Choanocytes are found almost exclusively in chambers,
rarely singly. The chambers are confined to the choanosome
in the center of the mesenchyme. In my opinion, these two
features are characteristic of this species. The cells contain inclusions that stain by methods directed toward demon-

stration of acid mucopolysaccharides and carbohydrates, but characteristically do not contain glycogen. The basal nucleus is small and spherical.

Fiber cells occur around coarse spongin B fibers. They characteristically have cytoplasmic extensions which are acid mucopolysaccharide-positive. The nucleus displays either a nucleolus or chromatin granules.

The special cell type, or granular cell (Figs. 8, 9), is spherical, larger than the"nucleolate"cell (5.9-11.5-17.6 X 5.9-9.1-10.2 μm), and is abundant among the spongin fibers. There is a nucleus, which may or may not contain nucleoli. The cells are filled with large (1 to 2 μm) inclusions which are acid mucopolysaccharide-negative, PAS-positive, and strongly glycogen-positive. These inclusions characterize the special cell type. RNA is not demonstrable in these inclusions.

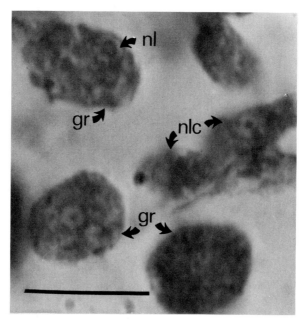

Fig. 8. "Nucleolate" cells (nlc) and granular cells (gr). A nucleus (nl) containing a nucleolus is visible in one of the granular cells. Bouin's, PAS-alcian blue. Scale 20 μm.

Spongin fibers are found throughout the mesenchyme. Both the reticular spongin A fibers and the coarser spongin B fibers are present.

226

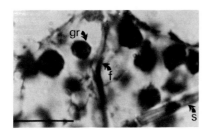

Fig. 9. Granular cells (gr) among spongin fibers (f). s-spicule. Bouin's, PAS-alcian blue. Scale 20 μm.

Callyspongia vaginalis (Lamarck, 1814, p. 436) de
   Laubenfels, 1936, p. 56.
= Spinosella vaginalis (Lamarck) Wiedenmayer, 1974,
   p. 367.

Morphology: Callyspongia vaginalis generally occurs as a
group of hollow tubes, but is sometimes found singly, and
partly or entirely solid. The diameter of the tubes increases distally. The atrium rim may be smooth or jagged.
The outer surface is tuberculate, and the inside of the
atrium is hispid, making it velvety to the touch. Oscules
occur along the interior of the tube and range in diameter
from 0.4 to 1.8 mm. The texture of the sponge is tough and
rubbery. The color is bluish-green, grayish-green, or gray-
ish-purple.
   The surface has the characteristic finer reticulating
spongin fibers among the coarser meshes. These fine fibers,
from 30 to 80 μm in diameter, radiate from the tips of the
conules. Interiorly, the fibers form circular and longitudinal tracts. The spicules are oxeas, slightly curved and
strongylote, which form plurispicular tracts in the spongin
fibers. The size of the spicules is 55.0-83.2-121.3 X 2.5-
3.1-4.5 μm.
   Cytology: "Nucleolate" cells, choanocytes, fiber cells,
and the special cell type, or granular cell, occur in
Callyspongia vaginalis. The "nucleolate" cells vary in size
and shape (4.4-7.0-10.1 X 4.4-5.4-7.3 μm) (Fig. 8). The
larger cells have a finely granular cytoplasm and the nucleolus is not often evident. The smaller cells are more
characteristic. The nucleus contains DNA, while the nucleolus and cytoplasmic granules are stained by methods for RNA.
The cytoplasm contains large spherical inclusions that stain
characteristically by methods for acid mucopolysaccharides
or glycogen deposits. These are probably vacuoles. Small,
basophilic cytoplasmic granules are also present.
   The choanocytes are found in chambers mainly in one re-
gion--the center of a cross-section of the wall of the tube,

227

the choanosome. Some are PAS-positive, glycogen-positive, or both. The nucleus is spherical and basal. DNA and RNA are present in the nucleus and cytoplasmic vacuoles.

Fiber cells are elongate, PAS-positive, and glycogen-negative. The nucleus contains a nucleolus or chromatin granules which stain for DNA. Cells are not grouped among reticular bundles. Rather, the coarse spongin B fibers predominate and the cells are scattered.

The special cell type is a granular cell (Figs. 8,9). The cytoplasm is finely granular and basophilic and is filled with larger, strongly PAS-positive and glycogen-positive inclusions. The nucleus may contain a nucleolus. Both the nucleus and cytoplasmic granules contain DNA and RNA.

DISCUSSION

Four cell types previously described were identified in the species studied: choanocytes, "nucleolate" cells, fiber cells, and pinacocytes.

Choanocytes are characteristic of the phylum. The choanocytes are about 3.5 μm in diameter, and chambers are approximately 20 μm in diameter. In all species studied, the nucleus is spherical and basal,and the cytoplasm contains carbohydrate and acid mucopolysaccharide inclusions. The occurrence of acid mucopolysaccharides in choanocytes has been reported in only the Haplosclerida and the Hadromerida (Stempien, '66); this occurrence, as well as the intensity of the staining reaction, may be diagnostic.

Acid mucopolysaccharides in sponges, analyzed previously by other workers (Stempien, '66; Garrone et al., '71) have indicated a relation to glycoproteins or mucin, rather than to the structural acid mucopolysaccharides such as chondroitin sulfates. Garrone et al. ('71) have suggested that acid mucopolysaccharides found in cell coats may be involved in ion exchange, cell aggregation, or collagen fibrillogenesis. I believe the strongly alcianophilic cytoplasmic inclusion is mucin which is secreted by the choanocytes and distributed throughout the sponge in the normal circulation of water to function as a mucous net to collect food particles or to provice a protective covering. One of the morphological characteristics of the order is the absence of dermal specialization, so the elaboration of a protective covering of mucus would be advantageous.

With the exception of Callyspongia plicifera, the choanocytes of all species contain glycogen. Whether the cells containing glycogen are storage or synthetic sites, however, is not known. Glycogen may also serve as an energy source to

support the needs of the beating flagella. In <u>Callyspongia</u>
<u>plicifera</u>, the choanocytes are PAS-positive and, therefore,
contain a non-glycogen carbohydrate. It is possible that
only ingestion but no digestion occurs in the choanocytes of
this species.

The precise mechanism by which food is ingested has not
been found, but it is known that the means of ingestion of
particulate food varies among the species, depending upon
the size of the food particles and the size of the choanocyte chambers (Jørgensen, '66).

In all species except <u>Callyspongia</u> <u>plicifera</u>, the choanocytes are found not only in chambers, but also scattered
throughout the mesenchyme. In <u>C</u>. <u>plicifera</u>, these cells
occur exclusively in chambers. In <u>Haliclona</u> <u>rubens</u>, <u>C</u>. <u>vagi-</u>
<u>nalis</u>, and <u>C</u>. <u>plicifera</u>, the choanocytes are generally locat-
ed in one area of the mesenchyme. This might be correlated
with the more orderly and less variable body form of these
three species.

"Nucleolate" cells occur in all species studied. These
cells are the classical amoebocytes described in the older
literature and some recent papers (see Pomponi, '74, for
review), and have been reported to function in ingestion,
digestion, storage of nutrients, and production of gametes.
They possess a prominent nucleolus in the nucleus, pseudopodia, and a variable number and kind of cytoplasmic inclusions which are eosinophilic, basophilic, acid mucopolysaccharide-positive, PAS-positive, or glycogen-positive. Most
cells contain vacuoles with or without stained granules.
Simpson ('63, '68b) concluded that the presence of the granules within vacuoles indicates ingestion or digestion of
food particles. "Nucleolate" cells are not localized, but
are often found in close proximity to choanocytes. In the
adult sponge, food particles ingested by the choanocytes
could be transferred to the neighboring "nucleolate" cells
to be digested and stored in the form of glycogen or glyco-
proteins. However, this process, particularly the manner in
which particles are disgorged from the cells, has not been
adequately described in the literature. The presence of acid
mucopolysaccharide containing inclusions indicates that the
"nucleolate" cells probably play a role in the elaboration
of the ground substance which completely surrounds them
(Pomponi, '74).

Simpson ('63) and Lévi ('56) found glycogen confined to a
single type of cell, the gray cell, in the species they
studied, and Lévi suggested that the Demospongiae have glycogen confined to this single category of cells. I have not
found this to be the case in the species I have examined.

All kinds of cells contain glycogen in varying degrees. It is likely that the capacity for glycogenesis is inherent in most cells of the sponge and that the monosaccharide precursors are available in the water that flows through the organism.

Fiber cells occur in all species studied. Only in Haliclona viridis, generally a fragile, amorphous, non-fibrous species, are these cells organized parallel to each other in bundles. In this species, the cellular bundles must assume a supporting role, since spongin is scarce. I believe the location and arrangement of the fiber cells and connective tissue is diagnostic, probably at the species level. In H. rubens, H. variabilis, and Dasychalina cyathina, the fiber cells are among thin spongin A fibers, which closely resemble vertebrate reticular fibers occurring in loose connective tissue. In Callyspongia vaginalis and C. plicifera, the thicker spongin B fibers are more abundant, and a few fiber cells occur near the spongin as well as scattered throughout the mesenchyme.

Special cell types, previously undescribed and possibly derivatives of "nucleolate" cells, occur in Haliclona variabilis, Dasychalina cyathina, Callyspongia plicifera, and C. vaginalis. Several large inclusions, 1 to 2 μm in diameter, fill the entire cell. These inclusions, perhaps twenty per cell, are strongly PAS- and glycogen-positive. A nucleus containing a nucleolus is often visible. The cells are spherical, and pseudopodia have not been observed. The granular cells are found almost exclusively in or bordering the spongin fibers. The histochemical reactions of the inclusions suggest that the cells function as reserves, but the similar staining affinities of the fibers and the granular cells, as well as the intimate association of the two, suggest an involvement in spongin synthesis. No previous workers have reported the occurrence of this cell type in relation to its probable role in spongin fibrillogenesis.

CONCLUSIONS

The arrangement and relative abundance of spongin fibers are diagnostic at the family level (Table 1). In the Haliclonidae, thinner spongin A fibers are more abundant than the wider spongin B fibers. There is generally no surface specialization of fibers, but some species, e.g., Haliclona variabilis and Dasychalina cyathina, have the tendency to produce a partial secondary reticulation. In the Callyspongiidae, spongin B fibers are predominant and are arranged in bundles, producing a more organized primary reticulation.

Secondary surface reticulation of fibers further character-
izes the family.

TABLE I

DIAGNOSTIC CHARACTERISTICS OF THE SPECIES STUDIED

Species	spongin	secondary surface reticulation	fiber cells	granular cells
Haliclona viridis	scarce	–	in bundles	–
Haliclona rubens	A > B	–	among spongin A	–
Haliclona variabilis	A > B	±	among spongin A	+ (few)
Dasychalina cyathina	A > B	±	among spongin A	++
Callyspongia plicifera	B > A	+	scattered	++
Callyspongia vaginalis	B > A	+	scattered	++

Choanocytes and "nucleolate" cells occur in all species
studied and cannot be used to separate the two families. The
presence of acid mucopolysaccharides in choanocytes, however,
may be diagnostic at the ordinal level.

Fiber cells occur in all species studied, but their location and arrangement are diagnostic, probably at the species
level. In Haliclona viridis, the fiber cells are organized
into bundles. In the other haliclonids studied, the fiber
cells occur among the spongin A fibers; in the callyspongiids
studied, the fiber cells are scattered throughout the mesenchyme.

The presence of special cells supports the separation of
the two families based on morphological characteristics
(Table I). It is difficult, however, to determine if the
granular cells are diagnostic at the family or generic level.
Haliclona variabilis and Dasychalina cyathina may be considered intermediate species, but present cytological find-

ings support biochemical and morphological observations and indicate that the Haliclonidae and Callyspongiidae should remain as two distinct families.

ACKNOWLEDGMENTS

This work was completed in partial fulfillment of the requirements for the degree of Master of Science. I express my appreciation to the members of my committee for their advice, support, and critical reading of the manuscript: Drs. C.E. Lane, D.P. deSylva, L.P. Thomas, D.L. Taylor, and Mr. R.C. Work. I also thank Susan S. Suarez for preparing the line drawings and Marilyn L. Taylor for typing the manuscript.

LITERATURE CITED

Bergmann, W. 1949 Comparative biochemical studies on the lipids of marine invertebrates, with special reference to the sterols. J. Mar. Res., 8(2): 137-176.

_____1962 Sterols: their structure and distribution. In: Comparative biochemistry. (M. Florkin and H.S. Mason, eds.) Academic Press, N.Y., 3: 103-162.

Bergquist, P.R. and W.D. Hartman 1969 Free amino acid patterns and the classification of the Demospongiae. Mar. Biol. (Berlin), 3(3): 247-268.

Bergquist, P.R. and J.J. Hogg 1969 Free amino acid patterns in Demospongiae, a biochemical approach to sponge classification. Cah. Biol. Mar., 10(2): 205.

Borojević, R. and C. Lévi 1964 Etude au microscope electronique des cellules de l'éponge: Ophlitaspongia seriata (Grant), au cours de la reorganisation apres dissociation. Z. Zellforsch., 64: 708-725.

Cheng, T.C., H. Yee, and I. Rifkin 1968a Studies on internal defense mechanisms of sponges: I. The cell types occurrin in the mesoglea of Terpios zeteki (de Laub.). Pac. Sci., 22(3): 395-401.

Cheng, T.C., H. Yee, I. Rifkin, and Kramer 1968b Studies on internal defense mechanisms of sponges: III. Cellular reactions in Terpios zeteki to implanted heterologous biological materials. J. Inv. Path., 12(1): 29-35.

de Laubenfels, M.W. 1932 Physiology and morphology of Porifera exemplified by Iotrochota birotulata Higgin. Carnegie Inst. Wash. Publ. No. 435, Pap. Tortugas Lab., 28: 38-66.

_____1936 A discussion of the sponge fauna of the Dry Tortugas in particular and the West Indies in general, with material for a revision of the families and orders of the Porifera. Carnegie Inst. Wash. Publ. No. 467, Pap. Tortugas Lab., 30, 225 p.

_____1950 The Porifera of the Bermuda Archipelago. Trans. Zool. Soc. London, 27(1), 154 p.

Dendy, A. 1890 Observations on the West-Indian chalinine sponges, with descriptions of new species. Trans. Zool. Soc. London, 12(10): 349-368.

Duchassaing, P. de Fonbressin and G. Michelotti 1864 Spongiaires de la mer Caraibe. Memoire publié par la Société hollandaise des science à Harlem. Natuurk. Verh. holland. Maatsch. Wet. Haarlem, 21: 1-124.

Fauré-Fremiet, E. 1931 Etude histologique de Ficulina ficus (Demospongiae). Arch. Anat. Micr., 27: 421-448.

Fell, P. 1969 The involvement of nurse cells in oögenesis and embryonic development in the marine sponge, Haliclona ecbasis. J. Morph., 127(2): 133-150.

_____1970 The natural history of Haliclona ecbasis de Laubenfels, a siliceous sponge of California. Pac. Sci., 24(3): 381-386.

Flax, M. and M. Himes 1952 Microspectrophotometric analysis of metachromatic staining of nucleic acids. Physiol. Zool., 25(4): 297-311.

Garrone, R. 1969a Une formation paracristalline d'ARN intranucleaire dans les choanocytes de l'eponge Haliclona rosea O.S. (Démosponge, Haploscléride). C.R. Acad. Sc. Paris, 269: 2219-2221.

_____1969b Collagène, spongine et squelette minéral chez l'éponge Haliclona rosea (O.S.) (Démosponge, Haploscléride). J. Microscopie, 8(5): 581-598.

_____1971 Fibrogenèse du collagène chez l'éponge Chondrosia reniformis Nardo (Démosponge, Tétractinellide). Ultra-structure et fonction des lophocytes. C.R. Acad. Sc. Paris, 273: 1832-1835.

Garrone, R., Y. Thiney and M. Pavans de Ceccatty 1971 Electron microscopy of a mucopolysaccharide cell coat in sponges. Experientia, 27: 1324-1326.

Harrison, F.W. 1972a The nature and role of the basal pinacoderm of Corvomeyenia carolinensis Harrison (Porifera: Spongillidae). Hydrobiologia, 39(4): 495-508.

_____1972b Phase contrast photomicrography of cellular behavior in spongillid porocytes (Porifera:Spongillidae). Hydrobiologia, 40(4): 513-517.

_____1974 Histology and histochemistry of developing outgrowths of Corvomeyenia carolinesis Harrison (Porifera: Spongillidae). J. Morph., 144(2): 185-194.

Harrison, F.W. and R.R. Cowden 1975 Cytochemical observa-

tions of larval development in Eunapius fragilis (Leidy):
Porifera; Spongillidae. J. Morph., 145(2): 125-142.

Harrison, F.W., D. Dunkelberger and N. Watabe 1974 Cytological definition of the poriferan stylocyte: a cell type
characterized by an intranuclear crystal. J. Morph.,
142(3): 265-276.

Hechtel, G.J. 1965 A systematic study of the Demospongiae
of Port Royal, Jamaica. Peabody Museum of Nat. Hist. Yale
Univ. Bull., 20, 103 p.

Jørgensen, C.G. 1966 Biology of suspension feeding. Perga-
mon Press, Oxford, p. 1-4.

Lamarck, J.B. 1814 Sur les Polypiers empatés. Ann. du
Muséum, 20: 370-386, 432-458.

Lévi, C. 1956 Étude des Halisarca de Roscoff. Embryologie
et systématique des Démosponges. Arch. Zool. Exp. Gen.,
93(1): 1-181.

_____1957 Ontogeny and systematics in sponges. Syst. Zool.,
6(4): 174-183.

Luna, L.G. ed. 1968 Manual of histologic staining methods
of the Armed Forces Institute of Pathology. 3rd ed.
McGraw-Hill Book Co., N.Y., 258 p.

Pallas, P.S. 1766 Elenchus zoophytorum hagae-comitum apud
petrum van Cleef.

Pavans de Ceccatty, M. 1957 La nature sécrétoire des
lophocytes de l'éponge siliceuse Chondrosia reniformis
Nardo. C.R. Acad. Sc. Paris, 244: 2103-2105.

Pavans de Ceccatty, M. and R. Garrone 1971 Fibrogenèse du
collagène chez l'éponge Chondrosia reniformis Nardo
(Démosponge, Tétractinellide). Origine et évolution des
lophocytes. C.R. Acad. Sc. Paris, 273: 1957-1959.

Pearse, A.G.E. 1968 Histochemistry: theoretical and
applied. 3rd ed. Vol. I. Little, Brown & Company,
Boston, 759 p.

Pomponi, S.A. 1974 A cytological study of the Haliclonidae
and the Callyspongiidae (Porifera, Demospongiae, Haplosclerida). M.S. thesis. Univ. of Miami, Coral Gables,
Florida, 90 p.

Schmidt, I. 1970 Phagocytose et pinocytose chez les
Spongillidae. Etude in vivo de l'ingestion de bactéries
et de protéines marquées à l'aide d'un colorant fluorescent en lumière ultra-violette. Z. vergl. Physiol., 66:
398-420.

Simpson, T.L. 1963 The biology of the marine sponge Microciona prolifera (Ellis and Solander). I. A study of cellular function and differentiation. J. exp. Zool., 154(1):
135-151.

_____1968a The biology of Microciona prolifera. II. Temperature-related annual changes in functional and reproductive elements with description of larval metamorphosis. J. exp. Mar. Biol. Ecol., 2: 252-277.

_____1968b The structure and function of sponge cells: new criteria for taxonomy of poecelosclerid sponges (Demospongiae). Peabody Mus. Nat. Hist. Yale Univ. Bull., 25, 141 p.

Stempien, M.F., Jr. 1966 The structure and location of sulfated and non-sulfated acid mucopolysaccharides as an aid to the taxonomy of Porifera: the class Demospongia. Amer. Zool., 6(3): 363 (abstract only).

Tessenow, W. Intracellular digestion of reserve substances in vacuoles of vesicular cells of Ephydatia muelleri. Naturwissenschaften, 55(6): 300.

Thiney, Y. 1972 Morphologie et cytochimie ultrastructurale de l'oscule d'Hippospongia communis Lmk. et de sa régénération. Thèse Univ. Claude Bernard, 63 p.

Tuzet, O. 1932 Recherches sur l'histologie des éponges: Reniera elegans (Bow.) et Reniera simulans (Johnston). Arch. Zool. Exp. Gen., 74: 169-192.

Tuzet, O. and J. Paris 1957 Les lophocytes de l'éponge siliceuse Tethya lyncurium Lamarck et leur évolution. C.R. Acad. Sc. Paris, 244: 3088-3090

Tuzet, O. and M. Pavans de Ceccatty 1953 Les lophocytes de l'éponge Pachymatisma johnstonni Bow. C.R. Acad. Sc. Paris, 237: 1447-1449.

_____1958 La spermatogenèse, l'ovogenèse, la fécondation et les premiers stades du développement d'Hippospongia communis Lmk. Bull.biol. Fr. Belg., 92: 331-348.

Tuzet, O., R. Garrone and M. Pavans de Ceccatty 1970 Origine choanocytaire de la lignée germinale male chez la Démosponge Aplysilla rosea Schulze (Dendroceratides). C.R. Acad. Sc. Paris, 270: 955-957.

Vacelet, J. 1967 Les cellules a inclusions de l'éponge cornée Verongia cavernicola Vacelet.´ J. Microscopie, 6(2): 237-240.

Wiedenmayer, F. 1974 Recent marine shallow-water sponges of the West Indies and the problem of speciation. Verhandl. Naturf. Ges. Basel, 84(1): 361-375.

_____(in press) A monograph of the shallow-water sponges of the western Bahamas.

Wilson, H.V. and J.T. Penney 1930 The regeneration of sponges (Microciona) from dissociated cells. J. exp. Zool., 56(1): 73-147.

ZOOGEOGRAPHY OF BRAZILIAN MARINE DEMOSPONGIAE

George J. Hechtel

Department of Ecology and Evolution
State University of New York at Stony Brook
Stony Brook, N.Y. 11794

ABSTRACT. The literature and Foster-Laborel collections
provide a basis for a preliminary zoogeographic study of
the Brazilian marine sponge fauna. The fauna exhibits a
strong affinity with that of the West Indies. Cosmotropical and provisionally endemic species constitute major
elements. There is little affinity with faunas of either
West Africa or the eastern Pacific. Distinctive faunas
may exist in the Recifan and Bahian regions of the coastline.

Brazilian sponges have received limited attention since the
nineteenth century Challenger Expedition. The present report is part of a study undertaken to expand our knowledge
of the alpha taxonomy and zoogeography of Brazilian marine
Demospongiae. At least 121 species of marine Demospongiae
have been reported previously from Brazil, excluding highly
probably synonyms. However, many records are based on beach-
worn and/or fragmentary specimens. New species and varieties in the nineteenth century literature often have inadequate descriptions. This is particularly true of keratose
species described by Hyatt (1877) and Poléjaeff (1884),
several of which are considered unrecognizable by de Laubenfels ('48). In total, at least 32 of the citations are based
on dubious identifications or brief descriptions, making
zoogeographic analysis difficult.

The Challenger Expedition collected 40 species of marine
Demospongiae along the northeastern bulge of the Brazilian
coast. Challenger stations were located near the Archipelago of Fernando de Noronha (03°50'S, 32°25'W), Recife in
Pernambuco, Barra Grande (09°09'S) and Maceio in Alagoas,
and Bahia. Most specimens were taken in shallow water (inshore to 25 fm = 45 m) on the coast of Bahia. A few were
dredged in deeper offshore waters (350-1715 fm = 630-3077 m)
between 09° and 10°S. The Challenger material was divided,
with typical Monaxonida described by Ridley and Dendy (1886,
1887), Tetractinellida and allied Monaxonida by Sollas (1886,

1888) and Keratosa by Polejaeff (1884).

Minor studies include those of Hyatt (1877) and Carter (1890), whose mostly questionable records are based largely on shore collections at Fernando de Noronha. Schultze (1865) described a <u>Darwinella</u> from Desterro, Brazil, for which the only modern listing is an inland town at $07°17'S$, $37°06'W$. Ridley (1881) reported a total of three species from the off-shore Hotspur and Victoria Banks ($17°32'S$, $35°46'W$; $20°42'S$, $37°27'W$). Selenka (1879) noted three species at Rio de Janeiro, including one previously reported from "Desterro" by Schmidt (1868). De Laubenfels ('56) provided a preliminary checklist, totalling 13 species, of shallow water Demospongiae collected near Camocin, in Ceará ($03°S$), Recife ($08°S$) and Santos ($24°S$). Mello Leitão et al ('61) provided a summary faunal list for all records through 1956.

In a major survey, Boury-Esnault ('73) recorded 60 species from the Brazilian collections of the Calypso. The Calypso stations included one at Fernando de Noronha and two at the nearby Atol das Rocas. An extensive series of coastal stations included nine near Recife ($07°29'S$ to $09°40'S$), eight in the Bahian region ($10°54'S$ to $16°47'S$), ten in the vicinity of Rio de Janeiro ($21°10's$ to $23°43'S$), two near Santos ($24°S$), and two in southern Brazil ($26°34'S$ and $32°41'S$). Five additional stations were located at or south of the Abrolhos Reefs, off southernmost Bahia. Most Calypso stations were at depths of 25-60m, although the overall depth range was 5-100 m.

My taxonomic study is based on the Brazilian sponge collections of R.W. Foster and Jacques Laborel, deposited at the Yale Peabody Museum, New Haven, Conn. Their combined collections include 2 partially and 52 completely identified species, of which 16 are new and 19 are new records for Brazilian waters (list 1). New genera and species will be described elsewhere (Hechtel, manuscript). The Foster-Laborel and Calypso collections have only nine species in common, despite overlap in study regions, attesting to the paucity of our knowledge and the diversity of the fauna. The total Brazilian faunal list of Demospongiae now totals at least 156 species.

Foster collected 68 specimens and fragments of Demospongiae, belonging to 23 species, during a molluscan survey in April, 1963. All specimens were taken at a single dredge station near Recife ($07°38.5'S$, $34°37'W$), in 15 fm = 27 m of water. Most specimens were attached to coral or other calcareous debris.

Laborel collected 85 specimens and fragments of Demospongiae, belonging to 45 species, during a survey of Brazilian coral reefs in 1961-64 (Laborel, '69). Most of Laborel's material was taken during SCUBA dives at 7-33 m, but some specimens were intertidal and others were dredged at 60-80 m. In the north, he obtained a few sponges at Fortaleza on the coast of Ceara, and in the Archipelago of Fernando de Noronha. Small collections were made toward the southern end of his reef study area, along the coast of Bahia, the Victoria Seamount, and the Abrolhos Archipelago (17°58'S, 38°30'W). Most of Laborel's specimens came from inshore coral banks along the coast of Pernambuco (about 08°S), particularly at or near Recife. A few specimens have no precise location due to disintegration of labels.

List 1. Systematic summary, Foster-Laborel collections. Notation: nr = new record for Brazil.

Class Demospongiae

Order Keratosa

Suborder Dictyoceratina

Family Dysideidae

  Ianthella ardis de Laubenfels (nr)

Family Spongiidae

Subfamily Spongiinae

  Hippospongia lachne (de Laubenfels)
  Ircinia fasciculata (Pallas)
  Ircinia strobilina (Lamarck)

Subfamily Verongiinae

  Fasciospongia caliculata (Lendenfeld) (nr)
  Psammaplysilla sp., cf. camera (de Laubenfels)
  Verongia fistularis (Pallas)
  Verongia longissima (Carter)
  Verongia sp. a
  Verongia sp. b

Order Haplosclerida

Family Adociidae

Adocia carbonaria (Lamarck)　(nr)
Adocia sp.

Family Callyspongiidae

Callyspongia sp.
Callyspongia pergamentacea (Ridley)
Callyspongia vaginalis (Lamarck)　(nr)

Family Desmacidonidae

Gelliodes ramosa (Carter)　(nr)
Fibularia (?) raphidifera (Topsent)　(nr) - requires new
　genus

Family Haliclonidae

Haliclona molitba de Laubenfels　(nr)
Prianos sp.

Order Poecilosclerida

Family Agelasidae

Agelas sparsus (Gray)

Family Anchinoidae

Anchinoe sp.
Echinodictyum sp.

Family Coelosphaeridae

Rhizochalina oleracea Schmidt　(nr)

Family Crellidae

new genus and species

Family Mycalidae

Monanchora barbadensis Hechtel　(nr)

Family Myxillidae

Iotrochota birotulata (Higgin)

Family Tedaniidae

   Tedania *ignis* (Duchassaing and Michelotti)

Order Halichondrida

Family Halichondriidae

   Halichondria sp.

Order Axinellida

Family Axinellidae

   Auletta sp.
   Axinella *lunaecharta* Ridley and Dendy (nr)
   Perissinella sp.
   Stylaxinella sp. a
   Stylaxinella sp. b
   Thrinacophora *funiformis* Ridley and Dendy

Order Hadromerida

Family Chondrillidae

   Chondrilla *nucula* Schmidt

Family Clionidae

   Cliona *celata* Grant

Family Latrunculidae

   Didiscus sp.

Family Placospongiidae

   Placospongia *carinata* (Bowerbank) (nr)

Family Spirastrellidae

   Spheciospongia sp.
   Spirastrella *cuspidifera* (Lamarck) s.s (nr)
   Spirastrella *(Anthosigmella)* *varians* (Duchassaing
     and Michelotti) (nr)
   Timea *mixta* (Topsent) (nr)

Family Suberitidae

  Aaptos bergmanni de Laubenfels

Family Tethyidae

  Tethya seychellensis (Wright) (nr)

Order Choristida

Family Geodiidae

  Erylus formosus Sollas
  Geodia (Geodia) gibberosa Lamarck
  Geodia (Cydonium) papyracea Hechtel (nr)

Family Stellettidae

  Rhabdastrella (Aurorella) sp.
  Scolopes moseleyi Sollas
  Stelletta (Pilochrota) anancora (Sollas) ss.
  Stelleta (Pilochrota) sp., cf. anancora (Sollas)

Order Spirophorida

Family Tetillidae

  Cinachyra (Cinachyrella) alloclada Uliczka (nr)
  Cinachyra (Cinachyrella) apion Uliczka (nr)
  Cinachyra (Cinachyrella) kukenthali Uliczka (nr)

Zoogeographers generally regard the coast of tropical Brazil as part of a tropical Atlantic American region (Ekman,
'53). Some treat tropical Brazil as a special subdivision,
distinguished from the West Indian by a component of endemic
Brazilian species. For example, Sollas (1888) reported endemic tetractinellid sponges from the Brazilian collections
of the Challenger Expedition. He then proposed a Brazilian
province extending southward from the Cabo de São Roque, in
Rio Grande do Norte, toward an undefined boundary with the
Magellanic fauna of Argentina.

The present study of marine Demospongiae supports inclusion
of Brazil within the tropical Atlantic American region, but
offers only limited support for recognition of a separate
Brazilian coastal province. The demosponge fauna is dominated by two zoogeographic components, namely, provisional
endemics and tropical Atlantic American endemics. Wide-

ranging species, with Atlantic and Indo-Pacific populations, may constitute a major component. However, records for many of these species require verification. Cosmotropical is used as a convenient term for such widespread species, but does not imply a continuous distribution. Indeed, most are absent from the tropical eastern Pacific. The amphi-Atlantic component of the Brazilian fauna is small, aside from cosmotropical species with eastern Atlantic populations.

Provisional Brazilian endemics account for 64 of the 156 species (42%) known from Brazilian waters (list 2). Of 52 Foster-Laborel species, 16 are new and four are previously-known Brazilian endemics (Thrinacophora funiformis, Scolopes moseleyi, Spirastrella cuspidifera s.s., and Stelletta anancora s.s.). The literature adds 44 Brazilian endemics, of which 29 are newly-described by Boury-Esnault ('73), and 6 are inadequately described by earlier authors. The high percentage of Brazilian endemics could reflect a zoogeographic reality, in view of the dispersal barriers to be discussed. In part, however, the apparent dominance of Brazilian endemics must reflect our limited knowledge of geographic ranges in tropical American sponges. As one caution, four (possibly five) of Sollas' endemic species now are known to have wider distributions. In addition, 19 of 36 previously-described Foster-Laborel species are new records for the entire Brazilian coastline. Boury-Esnault ('73) had 25 such records among 32 previously-described Calypso species. Furthermore, only a few localities have received detailed study in the adjacent West Indian region (Hechtel, '65). I employ the term West Indian for the warm water region north of 10°N, to expedite comparisons with the Brazilian region.

List 2. Provisional Brazilian endemics.
Notations on Brazilian records (used in all subsequent lists)

1-Boury-Esnault ('73), 2-Carter (1890), 3-de Laubenfels ('56), 4-Foster-Laborel collections, 5-Hyatt (1877), 6-Poléjaeff (1884), 7-Ridley (1881), 8-Ridley and Dendy (1886, 1887 or both), 9-Schmidt (1868), 10-Schultz (1865), 11-Selenka (1879), 12-Sollas (1886, 1888 or both, q = questionable identification, inadequate original description, or uncertain range.

Acanthacarnus radovani-1; Acarnus toxeata-1; Adocia sp.-4; Anchinoe sp.-4; Artemisina tylota-1; Auletta sp.-4; Basiectyon tenuis-8; Cacospongia amorpha-6, q; Cacospongia cincta-1; Cacospongia levis-6, q; Callyspongia coppingeri-7; Callyspongia sp.-4; Characella aspera-12; Clathria calypso-1;

Clathria raphida-1; species of Crellidae-4; Didiscus sp.-4;
Dysidea cana-5, q; Echinodictyum sp.-4; Ectoforcepia trilab-
is-1; Erylus corneus-1; Geodia neptuni-1,12; Geodia tylastra-
1; Geodia vosmaeri-1,12; Halichondria sp.-4; Holoxea violac-
ea-1; Ircinia compacta-6,q; Ircinia pauciarenaria-1; Jaspis
salvadori-1; Mycale fusca-8; Mycale quadripartita-1; Mycale
nuda-8; Paresperella spinosigma-1; Penares anisoxia-1; Penares chelotropa-1; Perissinella sp.-4; Placospongia cristata-
1; Polymastia corticata-8; Prianus sp.-4; Psammochela recife-
1; Psammochela tylota-1; Psammopemma porosum-6; Psammotoxa
nigra-1; Raphidophlus basiarenacea-1; Raspaxilla elegans-1;
Rhabdastrella sp.-4; Rhabdastrella virgula-1; Scolopes mos-
eleyi-4,12; Spheciospongia sp.-4; Spirastrella cuspidifera
s.s-4; Spongia bresiliana-1; Stelletta anancora s.s-1,4,12;
Stylaxinella sp. a-4; Stylaxinella sp.b-4; Suberites janeir-
ensis-1; Tetilla euplocamus-9,11,q; Tetilla radiata-11,q;
Thrinacophora funiformis-4,8; Timea agnani-1; Trachygellius
corticata-1; Trachytedania biraphidora-1; Verongia janusi-1;
Verongia sp.a-4; Verongia sp.b-4.

The Brazilian fauna of Demospongiae has pronounced affinities with that of the West Indian region (list 3). For example, 19 of the 36 previously-described Foster-Laborel
species are known only from Brazil and the West Indies, and
are thus tropical Atlantic American endemics. To reinforce
previous qualifications placed on known ranges, several
species are recorded from only one or two localities, often
far removed from Brazil. (As examples, Aaptos aaptos, Bermuda; Erylus formosus, Virgin Is.; Geodia papyracea, Jamaica;
Haliclona molitba, Bermuda, Bahamas; and Fibularia (?) raphidifera, Gulf of Mexico.) The literature provides 11 more
tropical Atlantic American endemics, of which three are in-
adequately-known Keratosa. An additional indication of West
Indian-Brazilian faunal affinity is provided by the occurrence of 21 cosmotropical and 7 amphi-Atlantic species in both
regions (although independent acquisition cannot be preclu-
ded). Finally, almost half of the new Foster-Laborel species
have significant similarities to West Indian ones.

List 3.   Tropical Atlantic American endemics
Aaptos bergmanni-1 (as A. aaptos),4; Agelas sparsus-1 (as A.
dispar),4; Axinella reticulata-8; Callyspongia vaginalis-4;
Caminus sphaeroconia-12; Cinachyra apion-4; Cinachyra kuken-
thali-4; Clathria calla-1; Corallistes typus-12; Dysidea
dubia-5,q; Erylus formosus-1,4,12; Fasciospongia caliculata-
4; Fibularia (?) raphidifera-4; Gelliodes ramosa-4; Geodia
papyracea-4; Haliclona erina-3; Haliclona molitba-4; Hippo-

spongia lachne-3,4; Ianthella ardis-4; Igernella joyeuxi-1;
Iotrochota birotulata-1 (as I. bistylata),4; Ircinia campana-
1; Ircinia longispina-6,q; Monanchora barbadensis-4; Rhizochalina oleracea-4; Spirastrella varians-4; Tribrachium
schmidtii-12; Verongia fistularis-3,4,6 (as V. tenuissima);
Verongia fulva-1, q; Verongia longissima-3,4

The occurrence of numerous Demospongiae in both the West
Indies and Brazil poses problems of dispersal routes and
mechanisms. Coastal regions between the mouths of the Orinoco and Amazon Rivers are generally unsuitable for hard bottom epifaunal organisms. In addition, the northwestwardly
flowing Guiana Current should hinder or prevent southward
passage of dissemules (rafters, larvae) from the West Indian
region. River outflows could affect the fauna by reducing
salinity, increasing turbidity, and depositing fine sediment.
In Brazil, Amazon river water lowers salinity over most of
the adjacent shelf, but effects decrease rapidly offshore
and at depths greater than 10 m (Magliocca, '71). Deposition
of fine sediment affects a much more extensive region, particularly in a northwest direction toward the Guianas (Laborel, '69).

The Brazilian coral fauna clearly is affected by these currents and river-associated barriers. Laborel ('69) emphasized the role of northwestwardly flowing currents in reducing
southward movement of coral larvae from highly diverse Caribbean reef communities toward Brazil. Corals are uncommon
to rare along the muddy-sand coasts from the Guianas to the
Cabo de São Roque in Rio Grande do Norte. Indeed, well-
developed reefs only commence south of the Cape (Laborel,
'69). Brazil is a peripheral zoogeographic region for corals, with a depauperate assemblage of West Indian genera and
species (Rathbun, 1879; Laborel, '67; Stehli and Wells, '71).
Endemic species are most evident in the fauna of well-illum-
inated shallow water habitats (Laborel, '69). West Indian
species with similar habitat preferences would be those most
affected by coastal barriers to dispersal and settlement.

Shallow water marine sponges also should be affected by
coastal barriers and currents. Few marine sponges extend
into brackish water (de Laubenfels, '47, 50a), and many are
affected adversely by heavy silting (Bakus, '68). Most Demospongiae require firm substrata, and many characteristic
West Indian species are associated with patch reefs and coral
rubble. I am unaware of any published records of marine
Demospongiae between the mouths of the Orinoco and Amazon
Rivers. West Indian species do occur in the mangrove swamps
of Trinidad (Hechtel, unpublished). South to north disper-

sal, aided by the Guiana Current, could be more significant
for sponges than for corals. The sponge faunas of Recife
and Bahia are certainly as diverse as those in individual
West Indian communities. Amphi-Atlantic and cosmotropical
species might spread to the West Indies via Brazil (see below). However, the West Indies probably exceeds Brazil in
total number of potential emigrant species, as shown by de
Laubenfels' incomplete checklist ('50a).

Sponges could traverse the Orinoco to Amazon barrier by
entrance of long-lived dissemules into northwestwardly-
bound ocean currents, or by step-wise colonization of fre-
quently-available shelf "oases" in either direction. Burton
('30, '32) regarded ocean surface currents as providing major
routes for sponge dispersal, with coastal creeping as a minor
factor. For example, the North Atlantic surface circulation
provided him with an apparent explanation for the affinities
between sponge faunas of Norway and the Azores.

The nature of such current-aided sponge dissemules remains
uncertain. Larvae of marine Demospongiae have only a limited
potential for long-distance transport and thus barrier-crossing. Sexual and asexual larvae are generally benthic creep-
ers or feeble swimmers with a mobile life measureable in
hours or days. Subtidal species of Polymastia do have benthic larvae that remain motile for several weeks (Borojevic,
'67; Bergquist and Sinclair, '68; Bergquist, Sinclair and
Hogg, '70). As an example of the difficulty, the distance
between Barbados W.I. and Natal, in Rio Grande do Norte is
about 3440 km. At a current speed of 2.3 km/hr. (Scheltema,
'68, for the South Equatorial Current), northward transport
would require about nine weeks, a duration that is beyond the
known survival time of sponge larvae.

Burton ('32) suggested that the dissemules might be larvae,
but were more likely newly settled rafters, carried along
with their substratum in the currents. Artificial rafting
by settlement on ship hulls, followed by release of young,
could be a significant means of barrier-crossing in historic
times. Ship-borne individuals are a suspected source of
West Indian and western Pacific Demospongiae in Hawaiian
fouling communities (de Laubenfels, '50b; Bergquist, '67).

Coastal creeping remains a possible means of passage, particularly for southward spread by West Indian Demospongiae.
Sediment studies by Ottmann ('59) suggest the presence of extensive oases of substrata suitable for Demospongiae (but not
corals) in northern Brazil. Ottmann found that fluviatile
deposits off the mouths of the Amazon were succeeded by an
area of clay-rich sediments, about 200 x 125 km in extent,
at a distance of 110 km off the Ilha de Marajo. A band of

shelly-sand deposits arced around the clay pocket and para-
lleled the coast for over 600 km, from the equator to 04°N.
An additional shell-sand station occurred between 05° and
06°N, off French Guiana (Ottmann, '59). Depths were under
100 m in the clay zone, and ranged from 30 - 108 m at shelly-
sand stations. Spicules of siliceous sponges occurred in
some sediment samples of each offshore zone. Laborel ('69)
suggested that the clay region was rich in sponges, and that
the shelly-sand region harbored a rich fauna of sponges,
bryozoans, and octocorals. If these suggestions are verified
shallow-water marine sponges could cross the Orinoco to Amazon barrier with short-lived rafters and/or larvae.

Amphi-Atlantic endemics are a minor component of shallow
water sponge faunas in both the West Indies (Hechtel, '65,
'69) and Brazil (list 4). Only 3 of 36 previously-known
Foster-Laborel species are found on both sides of the Atlantic, but not in the Indian or western Pacific Oceans. Axinella lunaecharta occurs off West Africa and the Cape Verde
Islands, Timea mixta in the Cape Verdes and Mediterranean-
Atlantic, and Cinachyra alloclada off the West African coast.
A fourth species, Geodia gibberosa, is recorded from the
Pacific coast of Panama, in addition to the Atlantic. Thenea
fenestrata, reported from Brazil by Sollas (1888), has an
equally extensive range, but occurs in deep water. The literature adds 13 more species to the list of potential amphi-
Atlantic species. However, six require confirmation for Bra-
zil, including three otherwise unknown in tropical Atlantic
America, and two need verification for the eastern Atlantic.
Uncertainty centers on the otherwise boreal species recorded
from Brazil by Ridley and Dendy (1887) and Carter (1890).

List 4. Amphi-American species
Axinella lunaecharta-4; Cinachyra alloclada-4; Cliona car-
teri-7, q; Craniella cranium-2, q; Geodia gibberosa-3,4;
Hymeniacidon perlevis-2 (as H. sanguinea), q; Oceanapia at-
lantica-8 (as O. robusta); Phakellia ventilabrum-8, q; Pronax plumosa-8, q; Spongia virgultosa-1; Stelletta crassi-
spicula-12; Stylocordyla stipitata-8, q; Tethya maza-11, 2;
Thenea fenestrata-12; Timea mixta-4; Timea stellifasciata-1;
Verongia capensis-3, q.

Although West Africa has a generally depauperate fauna
(Ekman, '53), numerous demosponges have been recorded from
coastal waters (Burton, '56; Levi, '59), and especially from
eastern Atlantic island groups such as the Azores (Topsent,
'28). However, the paucity of amphi-Atlantic sponges is
confirmed by Burton's West African study ('56), in which only

16% (9 of 58) of the Demospongiae collected by the Atlantide
Expedition had populations on both sides of the Atlantic.
My preliminary literature review of shallow water West Af-
rican sponges suggests that less than 30% of the species (in-
cluding cosmotropicals) occur anywhere in tropical Atlantic
America, and that less than 10% are amphi-Atlantic endemics.

Most West African and Brazilian sponges are probably steno-
topes, limited to warm shallow waters. The Atlantic basins
are a significant barrier for such organisms (Briggs,'67),
and probably have been since the end of the Cretaceous (see
Funnell and Smith, '68; Reymont and Tait, '72). Aside from
rafters, the barrier is crossed by some large motile species
(Briggs, '67) and by teleplanic invertebrate larvae (Scheltema, '68, '71). The latter are carried from west to east
either by the North Atlantic Drift or by a subsurface Equatorial Countercurrent. East to west passage is afforded by
the North and South Equatorial Currents. The minimum time
required for largely passive transport is measured in weeks
or months (Scheltema, '71). Sponge passage probably is restricted to infrequent rafters rather than larvae.

Amphi-American distributions are rare, with de Laubenfels'
Santos record ('56) of the Californian _Timea_ _authia_ requiring
verification. Only two otherwise amphi-Atlantic and four
cosmotropical Brazilian species extend into the tropical
eastern Pacific (list 5), which is an isolated region zoogeo-
graphically (Ekman, '53; Briggs, '67).

List 5. Species occurring in tropical Pacific America

Geodia _gibberosa_, _Lissodendoryx_ _isodictyalis_, _Placospongia_
carinata, _Spirastrella_ _coccinea_, _Tethya_ _diploderma_, _Thenea_
fenestrata, _Timea_ _authia_.

The Brazilian fauna of Demospongiae, like that of the West
Indies, has a component of cosmotropical, in some cases nearly circumtropical species (see lists 6 and 7). As many as
39 Brazilian species may belong to this category, including
9 in the Foster-Laborel collections. In addition, about 1/3
of the new species in this report have possible Indo-Pacific
allies. However, Brazilian identifications are uncertain
for at least 13 cosmotropical species reported in the literature, 8 of which are otherwise unknown in tropical Atlantic
America. For example, Ridley and Dendy (1888) reported six
Indo-Pacific species from Brazil on the basis of small specimens, fragments, or aberrant "varieties". In addition,
taxonomic problems obscure the geographical ranges of _Cran-
iella_ _carteri_ and at least three Foster-Laborel species (_Ir-

cinia fasciculata, I. strobilina, and Tedania ignis). All
may be conspecific with populations of similar sponges in
the eastern Atlantic and Indo-Pacific (Hechtel, '65).

List 6. Non-disjunct cosmotropical species

Azorica pfeifferae-12; Chondrilla nucula-1,2,3,4; Cliona
carpenteri-1; Cliona celata-1,4; Cliona schmidtii-1; Craniella carteri-12, q; Desmapsamma anchorata-8 (as Desmacidon
reptans); Dysidea avara-1; Dysidea fragilis-1,6 (as Spongelia
pallescens); Hippospongia equina-5, q; Ircinia fasciculata-
3,4, q; Ircinia strobilina-1,4, q; Lissodendoryx isodictyal-
is-3; Plakortis simplex-1; Plakina trilopha-1; Rhizochalina
fistulosa-8, q; Samus anonymus-12; Spirastrella coccinea-2
(as Chondrilla phyllodes), -3 (as Spirastrella cunctatrix);
Spongia officinalis-5,6, q; Spongia vermiculata-5, q; Suberites carnosus-1,2 (as S. massa), 8, q; Tedania ignis-1 (as
T. anhelans), 4, q; Tethya aurantium-2 (as Donatia lyncurium)
q; Tethya diploderma-3, q.

List 7. Disjunct cosmotropical species

Adocia carbonaria-4; Asteropus simplex-1 Axinella echidnaea-
8, q; Callyspongia fibrosa-8, q; Callyspongia pergamentacea-
4,7; Callyspongia robusta-8, q; Clathria procera-8 (as Rhaphidophlus gracilis), q; Darwinella mulleri-10; Holopsamma ar-
enifera-2, q; Ircinia ramosa-1; Placospongia carinata-4;
Rhizochalina putridosa-8, q; Tethya japonica-1; Tethya sey-
chellensis-4; Zygomycale parishii-3.

The remaining 22 Brazilian species (and additional West Indian ones) apparently do occur on widely separate tropical
coasts. They require explanation in terms of dispersal
routes, in view of the formidable intervening barriers of
temperature, land and deep water reviewed by Briggs ('67).
Of the 22, 5 may be unusually eurythermal, since Cliona celata, Dysidea fragilis and Tethya japonica extend into cold
temperate waters, and Plakina trilopha is known from the
Antarctic (Koltun, '66). Azorica pfeifferae is eurybathic,
which should reduce the effects of coastal and deep basin
barriers. The remaining species under consideration are presumably warm, shallow-water stenotherms, although experimental evidence is lacking. The extensive range of a few species may be based on ship-rafting. For example, Zygomycale
parishii occurs in previously-mentioned Hawaiian fouling
communities.
  Burton ('32) suggested a spread of cosmotropical species

from the diverse Indo-Pacific sponge fauna around the tip of South Africa into the Atlantic. Dissemules (rafters?) then could enter the northward flowing (cold water) Benguela Current and be carried to Brazil along the northern arc of the South Atlantic Gyre. Successful colonizers in turn could spread to the West Indies via the northward-flowing branch of the South Equatorial Current. The west coast of South Africa is an unlikely haven for tropical sponges, although current patterns are complex. Most tropical littoral invertebrates extend no further than East London on the east coast, although a few pass the Cape of Good Hope (Stephenson and Stephenson, '72). Sponges apparently exhibit a similar pattern, with only a minority of tropical species reaching the west coast (Lévi, '63). A few Indo-Pacific sponges do reach the Vema Seamount, at 31°38'S, 08°20'E (Lévi, '69).

Reid ('67) noted the difficulties of a South African route, particularly for warm water stenotherms. He explained the present multi-ocean distributions of some Hexactinellid genera on the basis of a Cretaceous Tethyan Seaway between the Indo-Pacific and Atlantic Oceans. Ekman ('53) invoked a similar explanation to account for generic similarities in the recent tropical Atlantic and Indo-Pacific faunas of several invertebrate groups, including crabs and echinoderms. Only extremely conservative species, as opposed to genera, could have a recent distribution based on a Tethyan dispersal. However, Wiedenmayer ('74) has supported a Tethyan ancestry for many recent species in his analysis of sponge speciation.

A curiously disjunct distribution exists for 9 of the 22 unquestioned cosmotropical Brazilian demosponges, which have populations in the Indo-Pacific and western Atlantic, but not in the eastern Atlantic (list 7). Similar patterns are noted for West Indian sponges by Wiedenmayer ('74). Such distributions, again on a generic rather than species level, have been attributed to a Tethyan dispersal, followed by late Tertiary climatic deterioration in the eastern Atlantic (Ekman, '53). Reid ('67) did demonstrate the existence of European Cretaceous species for some Hexactinellid genera with presently disjunct distributions.

The sponge faunas of tropical Brazil and cold temperate Argentina have little in common at the species level. Temperature differences and the outflow of the Rio de la Plata could act as faunal barriers (see Burton, '32; Knox, '60; Forest, '66). Most Argentine sponges are Magellanic or Antarctic in distribution (Burton, '32, '40). Of 35 Argentine species collected north of 40°S by Burton ('40), only 3 have been reported from tropical American waters. <u>Callyspongia</u>

pergamentacea, taken off Mar del Plata, in Buenos Aires province, is known from the Brazilian Hotspur Banks at $17^{O}S$ (Ridley, 1881) and from the Abrolhos Archipelago at $18^{O}S$ (present study). Hymeniacidon sanguinea, a synonym of H. perlevis, was reported doubtfully from Fernando de Noronha by Carter (1890). The third species, Stylohalina hirta, requires verification for Argentina. It is known from the West Indian and West African regions, but not from Brazil. Boury-Esnault ('73) recorded four Antarctic and/or Magellanic species from stations as far north as Rio de Janeiro (list 8). They may have lived in cold water areas and did not co-occur with any sponges known from clearly tropical regions.

List 8. Antarctic/Magellanic species

Ectyomyxilla kerguelensis-1; Suberites caminatus-1; Tedania murdochi-1; Tedania vanhoffeni-1.

Successive regions along the extensive, heterogeneous coastline of Brazil might differ significantly in their faunal composition. As examples, tropical Atlantic American Demospongiae might become less numerous toward the south, or coastal regions might harbor their own endemic species. Laborel's reef survey regions ('69) provide a useful means of comparison for shallow water sponges. His regions are delineated by differences in abundance of reef corals, and some by gaps between series of coral banks. Reef communities provide suitable habitats for many species of Demospongiae. In addition, his inter-regional gaps are associated with river outflows and sandy shores, which are generally unsuitable for sponge colonization, and potential barriers to dispersal.

Analysis is hampered by inadequate information, since most Brazilian Demospongiae are known from single localities. Only the coasts of Pernambuco and Bahia have received more than cursory attention. Apparent regional differences could reflect zoogeographic realities, or differences in sampling methods or sampled habitats. The present review clearly is preliminary, but does delineate problems and suggest geographic areas for further study.

Only six species of Demospongiae are known from the coast of Ceará in the northern coral-poor region that extends southward to the Cabo de São Roque (list 9). An additional 19 species are recorded for the Archipelago of Fernando de Noronha and/or the Atol das Rocas (list 10). However, at least 11 of those records are based on questionable identifications or inadequately described species. There is no

evidence for a distinctive Brazilian zoogeographic region between the outlets of the Amazon and the Cabo de São Roque.
Of 14 unquestioned species, 11 are recorded for more southerly Brazilian regions, and 8, (including the cosmotropicals
Chondrilla nucula and Spirastrella coccinea), are known in
the West Indian region to the north.   Only Prianus sp., and
the recently described Holoxea violacea and Rhaphidophlus
basiarenacea Boury-Esnault are endemic to the region on present evidence.   Additional collections from Fernando de Noronha are needed to clarify nineteenth century work, and to
make a detailed comparison with West Indian sponge faunas of
reef communities.

List 9.   Sponges of Ceará

Agelas sparsus-4; Hippospongia lachne-3; Ircinia fasciculata-
3; Verongia capensis-3, q; Verongia fistularis-3; Verongia
longissima-3.

List 10.   Sponges of Fernando de Noronha, Atol das Rocas

Agelas sparsus-1 (as A. dispar), 4; Chondrilla nucula-2;
Craniella cranium-2, q; Dysidea cana-5, q; Dysidea dubia-5,
q; Dysidea fragilis-1; Hippospongia equina-5, q; Holopsamma
arenifera-2, q; Holoxea violacea-1; Hymeniacidon perlevis-2
(as H. sanguinea), q; Ircinia pauciarenaria-1; Prianus sp.-
4; Rhaphidophlus basiarenaria-1; Spirastrella coccinea-2 (as
Chondrilla phyllodes); Spongia officinalis-5, q; Spongia vermiculata-5, q; Spongia virgultosa-1; Suberites carnosus-2 (as
S. massa), 8, q; Tethya aurantium-2 (as Donatia lyncurium),
q; Verongia janusi-1.

A Recifan reef region extends, with short gaps, from the
Cabo de Sao Roque to the Rio São Francisco (Laborel, '69).
Sponges are known mainly from 07-08°S, through Foster's
dredge haul, Laborel's reef samples, and Boury-Esnault's
Calypso studies ('73).   The Foster-Laborel collections include 44 Recifan species, and the regional faunal list totals
76 species for waters under 60 m in depth (list 11).

List 11.   Sponges of Recife and vicinity

Aaptos bergmanni-1 (as A. aaptos), 4; Acanthacarnus radovani-1; Adocia carbonaria-4; Adocia sp.-4; Agelas sparsus-1
(as A. dispar), 4; Artemisina tylota-1; Asteropus simplex-1;
Auletta sp.-4; Axinella lunaecharta-4; Cacospongia cincta-1;
Callyspongia vaginalis-4; Chondrilla nucula-3,4; Cinachyra

alloclada-4; Cinachyra apion-4; Cinachyra kukenthali-4; Clathria calla-1; Cliona carpenteri-1; Cliona celata-4; Cliona
schmidti-1; Dysidea fragilis-1; Echinodictyum sp.-4; Erylus
corneus-1; Erylus formosus-1,4; Fasciospongia caliculata-4;
Fibularia (?) raphidifera-4; Gelliodes ramosa-4; Geodia gib-
berosa-3,4; Geodia neptuni-1,12; Geodia papyracea-4; Geodia
tylastra-1; Halichondria sp.-4; Haliclona molitba-4; Hippospongia lachne-4; Ianthella ardia-4; Igernella joyeuxi-1; Iotrochota birotulata-4; Ircinia campana-1; Ircinia fasciculata
-4; Ircinia pauciarenaria-1; Ircinia strobilina-4; Lissodendoryx isodictyalis-3; Monanchora barbadensis-4; Penares anis-
oxea-1; Penares chelotropa-1; Perissinella sp.-4; Placospongia cristata-1; Plakina trilopha-1; Psammochela recife-1; Psammotoxa nigra-1; Rhabdastrella sp.-4; Rhabdastrella virgula-1;
Rhizochalina oleracea-4; Scolopes moseleyi-4; Spheciospongia
sp.-4; Spirastrella coccinea-3 (as S. Cunctatrix); Spirastrella cuspidifera s.s.-4; Spirastrella varians-4; Spongia bres-
iliana-1; Spongia officinalis-5; Spongia virgultosa-1; Strel-
letta anancora-4; Stylaxinella sp. a-4; Stylaxinella sp. b-4;
Tedania ignis-1 (as T. anhelans),4; Tethya diploderma-3;
Tethya japonica-1; Tethya seychellensis-4; Thrinacophora fu-
niformis-4; Timea mixta-4; Timea stellifasciata-1; Trachygellius corticata-1; Verongia fistularis-4; Verongia fulva-1;
Verongia janusi-1; Verongia sp. a-4; Verongia sp. b-4.

Reefs are absent along the sandy coast of Sergipe, south of
the Rio São Francisco. An ensuing Bahian reef region extends
southward from about 12ºS to the boundary of Bahia and Espirito Santo. It includes the offshore coral banks of the Abrolhos Archipelago and Victoria Seamount. Brazilian corals
attain their maximum development along the Bahian coast, in
terms of both reef structure and species diversity (Laborel,
'69). The Bahian shallow water fauna of Demospongiae totals
56 species, with 8 collected by Laborel and the remainder
largely by the Alert, Challenger and Calypso Expeditions
(list 12).

List 12.   Sponges of Bahian region, including Abrolhos,
           Hotspur, and Victoria banks

Aaptos bergmanni-4; Acarnus toxeota-1; Agelas sparsus-1 (as
A. dispar),4; Anchinoe sp.-4; Axinella echidnaea-8; Axinella reticulata-8; Azorica pfeifferae-12; Basiectyon tenuis-
8; Cacospongia amorpha-6; Callyspongia coppingeri-7; Callyspongia fibrosa-8; Callyspongia pergamentacea-4,7; Callyspongia robusta-8; Callyspongia sp.-4; Caminus sphaeroconia-
12; Chondrilla nucula-1; Clathria calypso-1; Clathria pro-

cera-8 (as Rhaphidophlus gracilis); Cliona carteri-7; Craniella carteri-12; Desmapsamma anchorata-8 (as Desmacidon reptans); Didiscus sp.-4; Dysidea avara-1; Dysidea fragilis-1,
6 (as Spongelia pallescens); Ectoforcepia trilabis-1; Erylus
formosus-12; Geodia glariosa-12; Iotrochota birotulata-1 (as
I. bistylata); Ircinia compacta-6; Ircinia ramosa-1; Ircinia
strobilina-1; Jaspis salvadori-1; Mycale fusca-8; Mycale nu-
da-8; Mycale quadripartita-1; Oceanapia atlantica-8 (as O.
robusta); Plakortis simplex-1; Pronax plumosa-8; Psammochela
tylota-1; Psammopemma porosum-6; Rhizochalina fistulosa-8;
Rhizochalina putridosa-8; Samus anonymus-12; Scolopes mose-
leyi-12; Spongia bresiliana-1; Spongia officinalis-6; Spongia
virgultosa-1; Stelletta anancora-4,12; Stelletta crassispi-
cula-12; Stylocordyla stipitata-8; Tedania ignis-1 (as T. anhelans); Tribrachium schmidtii-12; Thrinacophora funiformis-
8; Verongia fistularis-4; Verongia longissima-4; Verongia
fulva-1.

There is tentative evidence of significant endemism in the
demosponge fauna of both the Recifan and Bahian regions. A
total of 25 species is endemic to the Recifan region, including Geodia neptuni, Spirastrella cuspidifera s.s., 11 new
species, and 12 species recently-described by Boury-Esnault
('73). A total of 16 species is provisionally endemic to
the Bahian region, including 3 new species and 6 established
by Boury-Esnault. Other Bahian endemics are Cacospongia a-
morpha, Callyspongia coppingeri, Ircinia compacta, Basiectyon
tenuis, Mycale fusca M nuda, and Psammopemma porosum.
    The two groups of endemic sponges suggest a discontinuity
between the faunas of the Recifan and Bahian reef regions.
Indeed, the two faunal lists have only 16 species in common.
Furthermore, tropical Atlantic American species decline in
both number (22 vs. 10) and percentage of total fauna (29 vs.
18%) between the two regions.
    The faunal discontinuity could reflect the effectiveness of
the Rio São Francisco and the sandy Sergipe coast as barriers
to coastal creeping by Demospongiae. However, previous qualifications must be recalled, particularly since three species
of tropical Atlantic American sponges are known from Brazil
only in Bahian waters (Axinella reticulata, Caminus sphaero-
conia, and Tribrachium schmidtii). The Brazilian current
certainly should provide a southward route for dissemules,
including rafters, and the coast of Sergipe does offer some
suitable if non-reef habitats for Demospongiae. Boury-
Esnault ('73) recorded seven species from a substratum of
muddy-sand and shells (Forest, '66) at 10°54'S, off northern
Sergipe. Four of the seven occur in both Recifan waters to

the north and Bahian waters to the south, suggesting a continuous distribution. Present information unfortunately does
not permit a detailed comparison of Recifan and Bahian sponges from similar habitats.

The tropical Atlantic American region has an imprecise
southern boundary near Rio de Janeiro (Ekman, '53). Well-
developed reefs end at the northern boundary of Espirito Santo, which has a largely sandy coast. Cold water upwellings
(Emilsson, '61) restrict a dwindling array of corals to oases
in bays between Rio de Janeiro and Santos (Laborel, '69).
Cabo Frio is generally regarded as the northern boundary of
a warm temperate South Brazilian province (Knox, '64; Tommasi, '65), which Laborel ('69) regards as analogous to the
Carolina coast of the United States. However some tropical
invertebrates extend southward to the mouth of the Plata
(Ekman, '53).

The known sponge fauna of Brazil, south of Cabo Frio, totals 25 species, of which 3 are poorly described and 2 dub-
iously identified (list 13). Only two tropical Atlantic
American sponges are known, <u>Verongia longissima</u> and <u>Haliclona
erina</u>, with the latter unrecorded elsewhere in Brazil. However, at least five additional cosmotropical or amphi-Atlantic species are common to the West Indies and southern Brazil, although independent acquisition cannot be ruled out.
Of these, <u>Zygomycale parishii</u> is unrecorded elsewhere in Brazil, but the other four are known in at least two study regions. The disjunct coastal distributions of several species could reflect scattered sites of successful settlement
by dissemules, relict populations, or inadequate sampling.
Distinctive components of the southern Brazilian sponge fauna
include six endemic species described by Boury-Esnault ('73)
and four Antarctic and/or Magellanic species recorded by the
same author. Additional collections between Santos and the
mouth of the Plata would be invaluable in tracing the pre-
sumed transitions from a tropical through a warm temperate
to a cold temperate fauna.

List 13. Sponges of the Rio de Janeiro, Santos, and south

<u>Chondrilla nucula</u>-3; <u>Cliona celata</u>-1; <u>Cliona raphida</u>-1; <u>Ecto-
myxilla kerguelensis</u>-1; <u>Geodia gibberosa</u>-3; <u>Geodia vosmaeri</u>-
1; <u>Haliclona erina</u>-3; <u>Lissodendoryx isodictyalis</u>-3; <u>Pare-
sperella spinosigma</u>-1; <u>Raspaxilla elegans</u>-1; <u>Suberites cam-
inatus</u>-1; <u>Suberites carnosus</u>-1; <u>Suberites janeirensis</u>-1;
<u>Spongia bresiliana</u>-1; <u>Tedanis murdochi</u>-1; <u>Tedania vanhoffeni</u>-
1; <u>Tethya diploderma</u>- 3; <u>Tethya maza</u>-11; <u>Tetilla euplocamus</u>-
11; <u>Tetilla radiata</u>-11; <u>Timea agnani</u>-1; <u>Timea authia</u>-3;

Trachytedania biraphidora-1; Verongia longissima-3; Zygomycale parishii-3.

Sponges collected by the Challenger Expedition at depths of 630-3077 m off the coasts of Pernambuco and Alagoa (09°-10°S) are treated here as a separate deeper water component of the fauna, presumably little influenced by coastal barriers. Burton ('30) found pronounced eurybathy, and no correlation between vertical and geographic ranges, in his study of Norwegian sponges. Earlier ('28) he regarded the 100 fm (180 m) line as a significant shallow-deep water boundary, with Keratosa largely above and Hexactinellida largely below that level. I assume that temperature changes with depth would be more significant for tropical stenotherms than for Norwegian sponges. Sponge faunas in Jamaica and Barbados certainly change markedly between wading depths and 60 m (Hechtel, '65, '69; Reiswig, '73), under the influence of a variety of factors, including sedimentation and wave action (Reiswig, '73). De Laubenfels ('50a) selected 50 m as a general lower limit for West Indian shallow water sponges.

The nine deeper water Brazilian species collected by the Challenger include four Brazilian endemic, three tropical Atlantic American, two amphi-Atlantic and no costotropical sponges (list 14). Only two species are eurybathic, if Verongia tenuissima is really a synonym of V. fistularis. The second species, Geodia vosmaeri, was reported by Boury-Esnault ('73) from 39 m at 22°S, at a possibly cold-water station. Present evidence does not support my expectation of extensive geographic ranges for the deeper water fauna, but instead suggests a high degree of endemism. Further study of this fauna should be rewarding.

List 14. Sponges from deep water

Cacospongia levis-6; Characella aspera-12; Corallistes typus-12; Geodia vosmaeri-12; Ircinia longispina-6; Phakellia ventilabrum-8; Polymastia corticata-8; Thenea fenestrata-12; Verongia fistularis-6 (as V. tenuissima).

ACKNOWLEDGMENTS

I would like to express my appreciation to Dr. Willard D. Hartman of the Yale Peabody Museum, who made the Foster-Laborel collections available for study. The project was supported by a grant-in-aid from the Graduate School at Stony Brook, and research grant #031-7102A from the Research Foundation of the State University of New York. This is

contribution #86 of the Ecology and Evolution program at
Stony Brook.

LITERATURE CITED

Bakus, G. 1968 Sedimentation and benthic invertebrates of
Fanning Island, central Pacific. Mar. Geol., 6:45-51.

Bergquist, P. 1967 Additions to the sponge fauna of the
Hawaiian Islands. Micronesia, 3:159-173.

Bergquist, P., and M. E. Sinclair 1968 The morphology and
behaviour of larvae of some intertidal sponges. N.Z.Jl.
Mar. Freshwat. Res., 2:426-437.

Bergquist, P., M. E. Sinclair, and J. J. Hogg 1970 Adaptation to intertidal existence; reproductive cycles and
larval behaviour in Demospongiae. In: The biology of the
Porifera. W. G. Fry, ed. Academic Press, New York. Symp.
Zool. Soc. (London), 25:247-270.

Borojevic, R. 1967 La ponte et le développement de Polymastia robusta. Cah. Biol. mar., VIII:1-6.

Boury-Esnault, N. 1973 Campagne de la "Calypso" au large
des côtes Atlantiques de l'Amérique du Sud. Spongiaires.
Résult. scient. Camp. Calypso, X:263-295.

Briggs, J. 1967 Relationship of the tropical shelf regions.
Stud. Trop. Oceanogr. Miami, 5:570-573.

Burton, M. 1928 A comparative study of the characteristics
of shallow-water and deep-sea sponges, with notes on their
external form and reproduction. J. Queckett micro. Club,
(2) XVI:49-71.

_____1930 Norwegian sponges from the Norman Collection.
Proc. Zool. Soc. Lond., 2:487-546.

_____1932 Sponges. Discovery Rep., VI:237-392.

_____1940 Las Esponjas marinas del Museo Argentino de
Ciencias naturales. An. Mus. argent. Cienc. nat.,
XL:95-121.

_____1956 The sponges of West Africa. Atlantide Rep.,
4:111-147.

Carter, H. J. 1890 On the zoology of Fernando Noronha. J.
Linn. Soc., Zool., XX:564-569.

de Laubenfels, M. W. 1947 Ecology of the sponges of a
brackish water environment, at Beaufort, N. C. Ecol.
Monogr., 17:31-46.

_____1948 The order Keratose of the phylum Porifera--a
monographic study. Occ. Pap. Allan Hancock Fdn.,
3:1-217.

_____1950a An ecological discussion of the sponges of
Bermuda. Trans. zool. Soc. Lond., 27:155-201.

_____1950b The sponges of Kaneohe Bay, Oahu. Pacif.

Sci., $\underline{IV}$:3-36.

_____1956 Preliminary discussion of the sponges of Brazil. Controces. Inst. oceanogr. Univ. S Paulo, $\underline{1}$:1-4.

Ekman, S. 1953 Zoogeography of the sea. Sidgwick and Jackson, London. 417 p.

Emilsson, I. 1961 The shelf and coastal waters off southern Brazil. Bolm. Inst. Oceanogr. São Paulo, $\underline{10}$:102-112.

Forest, J. 1966 Compagne de la "Calypso" au large des côtes Atlantiques de l'Amérique du Sud. I. Compte rendu et liste des stations. Annls. Inst. océanogr. Monaco, $\underline{44}$:330-350.

Funnell, B. M. and A. G. Smith 1968 Opening of the Atlantic Ocean. Nature, London, $\underline{219}$:1328-1333.

Hechtel, G. 1965 A systematic study of the Demospongiae of Port Royal, Jamaica. Bull. Peabody Mus. nat. Hist., $\underline{20}$:1-94.

_____1969 New species and records of shallow water Demospongiae from Barbados, West Indies. Postilla, $\underline{132}$:1-38.

Hyatt, A. 1877 Revision of the North American Porifera, with remarks upon foreign species II. Mem. Boston Soc. nat. Hist. $\underline{II}$:481-554.

Knox, G. A. 1960 Littoral ecology and biogeography of the southern oceans. Proc. R. Soc. Lond., B $\underline{152}$:557-623.

Koltun, V. M. 1966 Sponges of the Antarctic. I. Tetraxonida and Cornacuspongida. Biol. Rep. Soviet. Antarctic Exped., $\underline{2}$:6-133.

Laborel, J. 1967 A revised list of Brazilian scleractinian corals, and descriptions of a new species. Postilla, $\underline{107}$:1-14.

_____1969 Les peuplements de Madreporaires des côtes tropicale du Brésil. Ann. Univ. Abidjon, (E) $\underline{2}$:7-261.

Levi, C. 1959 Résultats scientifiques des campagnes de la "Calypso". Golfe de Guinée. Spongiaires. Annls. Inst. océanogr., Monaco, $\underline{37}$:115-141.

_____1963 Spongiaires d'Afrique du Sud. I. Poecilosclerides. Trans. R. Soc. S. Afr., $\underline{37}$:1-72.

_____1969 Spongiaires du Vema Seamount (Atlantique Sud). Bull. Mus. Hist. nat., Paris, (2) $\underline{41}$:952-973.

Magliocca, P. S. 1971 Some chemical aspects of the marine environment off the Amazon and Para Rivers, Brazil. Bolm. Inst. Oceanogr. São Paulo, $\underline{20}$:61-84.

Mello Leitão, A., A. F. Pego, and W. L. Lopes 1961 Poriferos assinalados no Brasil. Avulso Cent. Estud. zool. Univ. Bras., $\underline{10}$:1-29.

Ottmann, F. 1959 Estudo das amostras do fundo recolhidas pelo N.E. "Almirante Saldanha", na região da embocadura do Rio Amazonas. Trabhs Inst. Biol. mar. oceanogr. Univ.

Recife, 1:77-106.
Polejaeff, N. 1884 Report on the Keratosa collected by
H.M.S. Challenger, during the years 1873-1876. Rep.
Chall. Zool., XI:1-88.
Rathbun, R. 1879 Brazilian corals and coral reefs. Am.
Nat., 13:539-551.
Reid, R. E. H. 1967 Tethys and the zoogeography of some
modern and Mesozoic Porifera. In: Aspects of Tethyan bio-
geography. C. G. Adams and D. V. Ager eds. Systematics
Assoc., London. Syst. Assoc. Publ., 7:171-181.
Reiswig, H. 1973 Population dynamics of three Jamaican
Demospongiae. Bull. Mar. Sci., 23:191-226.
Reymont, R. and E. Tait 1972 Biostratigraphical dating of
the early history of the South Atlantic Ocean. Phil.
Trans., (B) 264:55-95.
Ridley, S. 1881 The survey of H.M.S. 'Alert'. XL. Spongi-
da. Proc. zool. Soc. Lond., 1881:107-141.
Ridley, S. and A. Dendy 1886 Preliminary report on the
Monaxonida collected by H.M.S. 'Challenger'. Ann. Mag.
nat. Hist., (5) XVIII:325-351.
_____1887 Report on the Monaxonida collected by H.M.S.
Challenger during the years 1873-1876. Rep. Chall.,
Zool., XX:1-275.
Scheltema, R. 1968 Dispersal of larvae by equatorial ocean
currents and its importance to the zoogeography of shoal
water tropical species. Nature, Lond., 217:1159-1162.
_____1971 The dispersal of shoal-water benthic inverte-
brate species over long distances by ocean currents, In:
Fourth European Marine Biology Symposium pp. 7-28. D. J.
Crisp ed. University Press, Cambridge.
Schmidt, E. O. 1868 Die Spongien der Küste von Algier.
Engelmann, Leipzig., 44 pp.
Schultze, M. 1865 Ueber ein Exemplar von Hyalonema sieboldi
aus Japan und einem Schwamm mit Nadeln aus Hornsubstanz.
Verh. naturh. Ver. preuss. Rheinl., XXII:6-7.
Selenka, E. 1879 Ueber einem Kieselschwamm von acht
strahligem bau, und uber entwicklung der Schwammknospen.
Z. wiss. Zool., XXXIII:467-476.
Sollas, W. J. 1886 Preliminary account of the Tetractin-
ellid sponges dredged by H.M.S. Challenger, 1872-1876. I.
The Choristida. Scient. Proc. R. Dubl. Soc., V:177-199.
_____1888 Report on the Tetractinellidae collected by
H.M.S. Challenger, during the years 1873-1876. Rep.
Chall., Zool., XXV:1-455.
Stehli, F. G. and J. W. Wells 1971 Diversity and age pat-
terns in hermatypic corals. Syst. Zool., 20:115-126.

Stephenson, T. A. and A. Stephenson 1972 Life between tide-marks on rocky shores. W. H. Freeman, San Francisco. 425 pp.

Tommasi, L. R. 1965 Faunistic provinces of the western South Atlantic littoral region. Acad. Brazil cienc., Rio, 37 (Suppl.):261-262.

Topsent, E. 1928 Spongiaires de l'Atlantique et de la Mediterranée, provenant des croisières du Prince Albert Ier de Monaco. Résult. Camp. scient. Prince Albert I, LXXIV:1-376.

Wiedenmayer, F. 1974 Recent marine shallow-water sponges of the West Indies and the problem of speciation. Verhandl. Naturf. Ges. Basel, 84:361-375.

ECOLOGICAL FACTORS CONTROLLING SPONGE DISTRIBUTION IN THE GULF OF MEXICO AND THE RESULTING ZONATION

John F. Storr

Department of Biology
State University of New York at Buffalo
Buffalo, New York 14214

ABSTRACT: Sponge diversity and abundance along the northwestern coast of Florida in the Gulf of Mexico was found to be controlled by a combination of a number of ecological factors. The rapid decline in mean low temperatures northward was of major importance to overall decline in diversity. Zone by zone, however, it was found that factors such as rock bar abundance, lower wave activity, and the presence of the influx of nutrients from rivers increased abundance and diversity. These factors augmented growth reproduction rates and sponge diversities. Limiting factors were excessive algal growth, which killed sponges; wide sandy areas which inhibited sponge distribution because of the limited life span of sponge larvae; and high sedimentation rates resulting from strong tidal or wave activity which depleted the energy of the sponges. From the sponge diversity it was possible to establish zones of sponge distribution which, interestingly enough, corresponded closely with the sponging grounds of the commercial sponges.

As a result of data obtained on three extensive diving survey trips along the west coast of Florida during a two-year period from September 1955 through July 1956, together with information obtained from commercial sponge divers, it has been possible to determine the factors governing the ecology of sponges in general on the Florida coast in the Gulf of Mexico, and to delineate the specific distribution of individual sponge species. A number of important ecological zones have become apparent as a consequence of this delineation.

The area surveyed extended from Ten Thousand Islands in the south northward to St. Mark's on the west coast of Florida. Forty-nine species of sponge were found and identified, and the distribution of each separate species was determined and plotted. When this information was consolidated, and the distribution for all significantly represented species in the surveyed area was plotted, nine distinct zones of distribu-

tion became apparent in the area from Tampa Bay to St.
Mark's. The boundaries of these zones were established by
1) obvious breaks or groupings in the distribution of particular species, and 2) the northern or southern distribution
limits of particular species.

The area surveyed for this study had previously been divided by the commercial spongers into small regions, or
'sponging grounds', to geographically locate an area in
which a sponging boat had been, or would be, working. Sponge
fishermen claim that they can identify, to within 10 miles,
the location from which a commercial wool sponge was taken,
and thus the 'sponging grounds' also serve to label areas
yielding sponges of varying and recognizable appearance,
texture, and elasticity. The commercial spongers' zonation
in this respect, however, is dependent upon the grade or
quality of the sponge, and particularly of the wool sponge
(Hippiospongia lachne) which accounts for approximately 90%
of their annual yield on a per piece basis, whereas the zonation presented in this report was derived from the DISTRI-
BUTION of individual species of both commercial and non-
commercial varieties, and was little dependent upon the distribution of wool sponges since they are widely and virtually
evenly distributed throughout the area. It was, therefore,
interesting and significant to find a strong correlation between the two zonation plots. An effort has been made to
determine the environmental and ecological factors responsible for the variations and the determined distribution of the
sponges growing in the surveyed area.[1]

METHODOLOGY

In three surveying field trips, 97 diving stations were
established, covering the area from Ten Thousand Islands to
Carrabelle on the Gulf coast of Florida. Over 100 dives were
made. On the first trip, in September 1955, the area was

---

[1]The U.S. Fish and Wildlife Service contracted with the
Marine Laboratory of the University of Miami, in 1955, to
carry out a scientific investigation of commercial sponges
of Florida. The investigations were financed with funds
made available under the Act of July 1, 1954 (68 Stat. 376),
commonly known as the Saltonstall-Kennedy Act. The principal results of the investigations are presented in the paper
'Ecology of the Gulf of Mexico Commercial Sponges and its
Relation to the Fishery' by John F. Storr, 1964. U.S. Fish
and Wildlife Service Special Scientific Report-Fisheries No.
466, Washington, D.C.

surveyed by the author from Tarpon Springs to St. Mark's in water depths from 6 to 60 feet. The diving was also done by the author on the second trip, in November 1955, when the area from Tarpon Springs southward to the Ten Thousand Islands was investigated. The surveying on the third field trip in July 1956 was done by two commercial sponge divers, who explored, under direct supervision, the area from Tampa Bay to Carrabelle to the 60 foot depth.

At each station, dives were made from an anchored boat. A 100-foot air hose was used by the author to assist in standardizing the amount of area surveyed at each station. All species of sponges and other bottom organisms encountered while circling the area were collected, preserved, and later identified. A written description of the bottom was included in the report from each station and the abundance of various sponge species was noted. Only those species occurring in relative abundance at each station were considered in delineating zones; stray specimens of species found well outside these normal ranges were disregarded.

All sponge species were identified in the laboratory by examination of spicules and other morphological features. Any species for which there was any degree of uncertainty were sent to Dr. Willard Hartman of the Peabody Museum of Natural History for confirmation of specification.

ECOLOGICAL CONSIDERATIONS OF THE SURVEYED AREA

Six significant ecological factors were found to affect the sponges growing within the surveyed area: 1) water depth, 2) bottom structure, 3) water currents, 4) temperature, 5) nutrient levels, 6) sedimentation.

Water depth. The diving stations for the study ranged in depth from 9 to 60 feet. Sponges in the shallower waters were subject to two types of environmental stress not encountered by those in deeper waters. First, heavy algal growths may break loose and roll along the bottom in large mats, covering and killing sponges. Second, a sudden freshening of the water in local areas, resulting from heavy rains and sudden runoffs from rivers, may cause a considerable loss of sponges (Rathbun, 1887; Moore, '10; Crawshay, '39; and Storr, '64).

There were, for example, noticeable changes in the appearance of wool sponges taken from water of less than 12 feet as compared to sponges taken from 40 or more feet in depth. Those from shallow water had up to twice as many conule tufts per square inch, and the oscules of the shallow water sponges were as much as $1\frac{1}{4}$ inches wide. Those of sponges from deeper water were rarely more, and usually less, than three-quarters of an inch in diameter. The 'chimneys' which sur-mount the oscules were normally taller in shallow water sponges. There were also noticeable differences in the com-pressibility and strength of the fibers: compared to deeper water sponges, the shallow-water wool sponges were often very soft and weak, reflecting the rapid growth typical of warmer waters. The microscopic fiber structures from shallow water sponges were finer and more loosely connected than those of sponges that had grown in deeper waters. All of these differences resulted from the more rapid growth of those sponges in shallow water. Wool sponges growing in shallow water are also much darker in color than those found in deeper water due to greater light interaction. Similarly, Spongia barbara is black, dark brown, or creamy tan in color, depending on the area and depth from which it was taken (Storr, '64).

Bottom structure. The Society of Economic Paleontologists and Mineralogists ('55) reported that in the area from Tampa Bay to Carrabelle the average bottom slope was 2.6 feet per mile. This slope was maintained to approximately 180 feet in depth, and comprised the inner shelf. From the shore to the deeper water, the unconsolidated sediments could be di-vided into a series of zones roughly parallel to the coast. The INSHORE ZONE is about 20 miles in width, and is composed of 50% quartz sand. These sediments also contain abundant phosphorite grains, as well as shell, coral fragments, cal-careous algae remains, and sponge spicules. Seaward of the inshore zone is a QUARTZ-SHELL ZONE, in which shell material replaces quartz in making up more than 50% of the sediment. Other constituents of the sediments are similar to those found inshore, with the shell and other sediments of marine origin becoming more predominant seaward. No quartz sand is found seaward of this zone, which also averages approximately 20 miles in width.

The sponges reported in this study were restricted entirely to the INSHORE, or more than 50% quartz sand area where they are found growing attached to the rock of the rocky limestone bars.

Clean, exposed rock is essential for the attachment of wool and other species of sponges. The limestone rock bars sup-porting sponge growth may range from a few square feet to many thousands of square yards. Many rock bars are at the same level as the surrounding bottom, or only a few inches higher. Some rock bars, kept free of sediment by wave and tidal activity, are found even as much as six feet below the level of the flat bottom. The highest rocky bar area extends in a northwest direction from Anclote Key to the Cedar Keys area and supports excellent sponge growth. The rock is rough in texture, and may be as much as 12 feet in height (Storr, '64).

The two sand bars located in the surveyed area, as shown in figure 1, were probably important barriers to the distribution of sponges.

Water currents. Water currents, as depicted in figure 1, were probably the most important ecological factor affecting the distribution of sponges in the surveyed area: the redistribution pattern of the wool sponge, decimated by disease, along the coast north of Tampa Bay coincided almost exactly with the prevailing currents. Furthermore, at stations where the current was strong, sponges grew more profusely, and shape and texture varied greatly from that of the same species found in zones with weaker currents.

The rate of flow is affected by bottom contour, and is modified considerably by the rate of flow of the inshore current which travels at a rate of at least one-third of a knot in water 40 feet deep. Mean tidal flow along most of the coast in water depths of 24 feet is about one-half knot. In other areas, however, such as at Indian Rocks, a current of two knots or more may exist at times, and may be in part responsible for the rapid growth of some sponges. The more or less elliptical excursion varied in the area, decreasing by half for every three-fold increase in water depth.

The only way an area can naturally become populated by most marine sponges is by the drifting of their planktonic larvae into the area. When mature, spermatozoa are released from sponges into the open water and are taken up by egg-producing sponges in their incurrent flow of water. Spermatozoa enter an egg-producing sponge by chance; thus, the number of spermatozoa taken in, and the number of larvae produced, is in direct relation to the concentration of sperm in the water. The released larvae live as a free-swimming form for a period of from a few hours to several days. The larvae must then come in contact with clean, hard bottom to develop successfully into mature sponges. Those that are poorly anchored are usually washed about and killed.

Compared to the strong inshore current flowing parallel to the shore, the back and forth movement of the tide appeared to be relatively ineffective in dispersing the larvae along the shore. Tidal excursion was seen to be important to the fertilization process, however, since it carried released sperm back and forth over the sponge-producing area, greatly increasing the chances of fertilization. Wave action plays a similar role on a much smaller but still important scale.

The offshore eddies were found to be important for both dispersing the sponges seaward, and for raising local nutrient levels by lifting bottom sediments. The three eddies between Tampa Bay and the Cedar Keys area were reported to be between 20 and 25 miles in their greatest diameter (Storr, '64). Estimated current velocity in these eddies was one mile per day. The most northerly eddy, located south of St. Mark's and about 50 miles in its greatest diameter, was found to be quite important to the distribution of sponges to deeper water areas of the Gulf.

Temperature. Temperature variation in the area from St. Mark's to Tampa Bay is considerable and of major importance in restricting the distribution of a number of species of sponges for a variety of reasons. The average surface mean temperature and mean low temperature are given for various areas on figure 1 and in the table below.

	A.M.T. ^{o}C ^{o}F	M.L.T. ^{o}C ^{o}F
St. Mark's to Piney Point	21.9(71.4)	13.6(56.5)
Piney Point to Cedar Keys	22.4(72.4)	15.4(59.7)
Cedar Keys to Tarpon Springs	23.2(73.5)	16.9(62.5)
Tarpon Springs to Tampa Bay	24.2(75.6)	17.9(65.3)
Shark River Area (South Florida)	26.4(79.6)	22.1(71.8)
Key West Area	25.8(78.5)	21.6(71.0)

The temperature data were obtained from U.S. Department of Commerce Special Publication No. 278 ('55), which gives the surface water temperatures at tide stations along the Atlantic and Gulf of Mexico coasts. Accuracy to within $3^{o}F$ is claimed, depending on the location of the tide station on the shore. The lower temperatures of the upper Gulf of Mexico along the shores of Florida are related to the very low slope of the bottom and the broad shallow areas. With such a limited water mass and poor circulation, the water quickly loses its heat in winter when air temperatures are low.

A significant number of species did not occur in the most northern zones of the studied area, while others showed definite southern limits. Variations in temperature affected the production and release of sperm, the receptiveness of the eggs for fertilization, the size of the sponge in general, and the size of the sponge at maturation (when larval production was possible) (Storr, '64). Thus, temperature was undoubtedly a major limiting factor affecting the distribution of sponge species.

The wool sponge (Hippiospongia lachne), for example, began larval production only when the mean monthly temperature rose to about 73°F, and increased rapidly until the temperature exceeded 84°F. With a rise of 2 or 3 degrees over 84°F, the number of sponges producing larvae fell from 70% or more to about 25%. Likewise, below 73°F, larval production ceased. In addition, the greater the number of months per year in which the temperature was greater than 80°F, the smaller the minimum size of mature wool sponges at maturation. Consequently, wool sponges mature at 2 inches in diameter in Honduras (one year of growth), but mature at 5½ inches in diameter in the northern Gulf at St. Mark's, about 4 years of growth.

A similar relationship may exist between size at maturation and temperature variation with water depth. Thus, although there is insufficient data available to plot temperature durations at various depths throughout the year, it is possible that the dispersion of wool and other sponges to deep water areas in the Gulf may be retarded by the effect of temperature on sexual maturation.

Prolonged low water temperatures found in the shallow, northern waters may kill sponges or drastically deter their growth. The mature sponges of several species are also killed by the sudden changes in water temperatures which occur in these areas when the rate of change is greater than 0.5°C per day.

Nutrient levels. If food intake falls below demands of the living sponge tissue, the larvae may not become vigorous enough to leave the sponge or, if they do leave, they may not be strong enough to attach and metamorphose into young sponges (Storr, '64). Rivers flowing into the Gulf of Mexico carry considerable amounts of nutrients necessary for sponge growth, and a correlation was seen between the location of rivers and the distribution and concentration of sponges. A high level of available nutrients is also important to sponge dispersion, since it has been found that the total volume of reproducing sponge material present in an area, and not the number of sponges alone, must be considered as one 11-inch sponge may produce more larvae than six 6-inch sponges.

For individual sponges of a number of species (Hippiospongia, Spheciospongia) lack of sufficient nutrients in the water may result in starvation of the central area of the sponge. In such cases the sponge becomes doughnut shaped as it continues growth.

Sedimentation. No effects of sediment deposition on
sponges have been measured. It has been observed, however,
that the greater the amounts of sediment present, the greater
the energy utilization in getting rid of these sediments, or
the greater number of canals plugged with sediments, reducing
circulation. Sediment piling upon rocky areas used for attachment also reduce the probability of larval attachment and
few sponges were ever found at the edges of the rocky bars
which periodically are covered by sediments.

RESULTS

Forty-nine species of sponge were found and identified
within the surveyed area. Of these, the following six
species were found to be widely and quite evenly distributed:

Geodia gibbersoa
Hippiospongia lachne
Homaxinella rudis
Ircinia fasciculata
Microciona juniperina
Spheciospongia vesparium

Even among these species, however, ecological considerations
produced zonal variations. For example, eggs were generally
not produced by the wool sponges (Hippiospongia lachne)
growing north of Cedar Keys until they were 5½ inches in
diameter, while in the Gulf north of Tampa Bay, the minimum
diameter of wool sponges producing larvae was 5 inches
(Storr, '64). In addition, previously discussed variations
in the appearance, texture, and elasticity of certain
species result from differing ecological conditions.

Thirteen of the identified species were not found at a
sufficient number of stations to be significant in the
determination of distribution zones:

Aplysilla sulfurea	Desmacella pumilio
Aulena columbia	Mycale angulosa
Fibulia massa	Higginsia strigilata
Haliclona rubens	Aaptos bergmanni
Haliclona subtriangularis	Placospongia melobesioides
Stelletta grubii	Tethya diploderma
Chondrilla nucula	

Although these and the evenly distributed species were not
important in the determination of the distributive zones

shown in figures 1 and 2, it was significant that certain
areas did support a greater total number of sponge species
than others, and this information is included on the accom-
panying chart.
Distribution of sponge species. Thirty species of sponge
were found to have significant environmental distributions.
These distributions were plotted, and their northern and,
when possible, their southern limits were determined. This
information yielded the plots shown in figure 2, along with
the suggested zonation. The numbers included in figure 2
refer to the following sponge species:

1. Cryptotethya crypta
2. Axinella polycapella
3. Cliona caribboea
4. Callyspongia vaginalis
5. Haliclona viridis
6. Lissodendoryx isodictyalis
7. Merriamium tortugaensis
8. Thalyseurypon vasiforme
9. Spongia graminea
10. Ircinia campana
11. Ircinia strobilina
12. Aplysina longissima
13. Darwinella mulleri
14. Dysidea crawshayi
15. Spongia officinalis
16. Spongia zimocca
17. Aplysina fistularis
18. Callyspongia arcesiosa
19. Callyspongia procumbens
20. Neopetrosia longleyi
21. Haliclona variabilis
22. Tedania ignis
23. Xytopsues griseus
24. Homaxinella waltonsmithi
25. Anthosigmella varians
26. Cliona lampa
27. Spirastrella coccinea
28. Cinachyra cavernosa
29. Myriastra debilis
30. Ianthella ardis

Arrows in figure 2 indicate that no southern limit was determined.
Each species was analyzed as to the zone(s) in which it was
found, and the depth of each station where it was found.
Little correlation between depth and specific species distribution was found; only one species, Spongia graminea,
was restricted to water less than 24 feet deep. Only four
species, Aplysilla sulfurea, Microciona juniperina, Haliclona
subtriangularis, and Haliclona variabilis were only found in
water more than 24 feet deep. However, shallow waters in
general were found to support only a limited number of
species. This may have been the additive result of 1)quickly
fluctuating water temperatures in response to changing air
temperatures, and 2) a general lack of active circulation
by water currents in shallow water areas. In some instances,
the few species that were found in a particular shallow
area were all more or less the same size (Storr, '64) indicating, perhaps, that larvae had been brought into the area
during one or two short periods when the current had swung

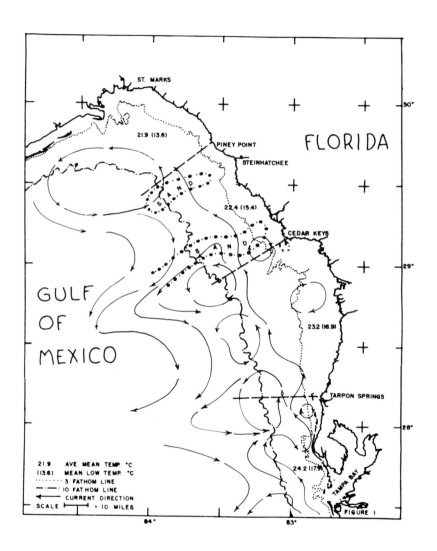

Fig. 1.

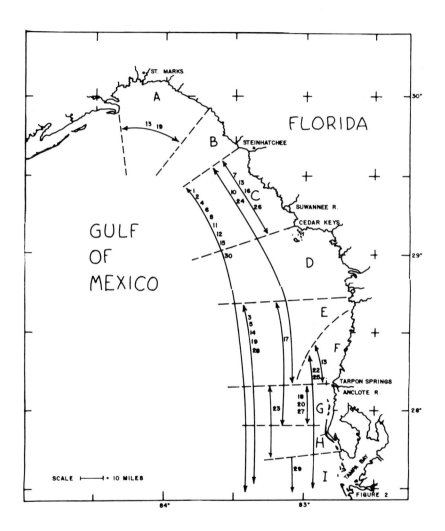

Fig. 2.

towards shore.

The density of growing sponges was not important to the zonation determined by species distribution as shown in figure 2. Zone D, for example, supported an extremely dense sponge population (Storr, '64), and yet relatively few species were represented in this area. It furthermore was found to be either a northern or a southern distribution limit for a significant number of species, as indicated in figure 2.

A discussion of the distribution zones.

ZONE A: This was the most northern zone of the surveyed area, with shallow, relatively calm waters, poor circulation, and a gently sloping bottom structure. Many sponge species were excluded from this zone, most probably due to the minimum temperatures encountered. The yearly mean temperature was reported by the U.S. Department of Commerce Special Publication No. 278 ('55) to be 71.4°F, with a lowest mean temperature of 56.5°F. Although most of the sponge species found in the surveyed area can grow within these ranges, they are well below the normal optimum ranges both for growth and particularly for reproduction. Furthermore, the temperature of the shallow water of this zone fluctuates with the air temperature.

The ten sponge species found in Zone A were concentrated near the mouth of the Aucilla River, where both nutrients and rock bars were plentiful.

ZONE B: The environmental conditions of Zone B were quite similar to those found in Zone A: bottom slope, water depths, yearly mean and lowest mean temperatures were comparable. Nevertheless, only three species were represented in this zone, and these were species that were found abundantly throughout the entire surveyed area. Distribution was probably limited not only by low, quickly varying temperatures and weak currents, but possibly by low nutrient levels also. As well, the wide sand bar area forming the southern boundary of this zone probably limited sponge distribution.

One other limiting factor may be present. In studying reproduction it was apparent that the concentration of a sponge species must reach a specific level before optimum reproduction levels were reached. The fewer bar areas and possibly the lower nutrients and fewer larval drift from the south may all have had an impact on the low numbers of sponge species.

ZONE C: Thirty-nine sponge species were found in this zone, which far exceeded the number found in any other zone. Of the thirty species having significant environmental distribution, 16 were found in Zone C. Environmental conditions in this area were diverse and generally favorable for both growth and reproduction of sponges. A yearly mean water surface temperature of 72.4°F, with a lowest mean temperature of 59.7°F was reported. Water temperatures changed gradually in response to lowered air temperatures. The bottom slopes were steeper than in A or B, and many rock bars were found in water depths varying from less than 20 feet to more than 50 feet. Although rather strong currents were located parallel to the shore, there was little turbulence   Two rivers, the Steinhatchee and the Suwannee flow into this zone, resulting in greater nutrient inputs. The large number of rocky bars and adequate nutrient supplies would result in optimum levels of concentrations for reproduction.

Large sand bars, acting as major barriers to the dispersion of sponges and nutrients, formed both the northern and southern boundaries of Zone C. As a probable consequence, this zone was the northern distribution limit for twelve of the significantly distributed species (see Distribution of sponge species.), and in addition, four species were found ONLY in this zone. The combined effects of shallower waters, lower mean temperatures, the termination of the steady, northerly inshore current (Fig. 1), and the physical barrier of the sand bar probably prevented the northerly dispersion of a significant number of species from Zone C into Zone B, where only three sponge species were represented.

ZONE D: Although established species grew well in this zone, it was delineated more by factors deterring the dispersion of sponges than by favorable environmental factors. Water temperatures were, in general, favorable for both growth and reproduction: reported yearly mean temperature was 73.5°F, with a lowest mean temperature of 62.5°F. Nevertheless, in the shallow waters of the southern section of this zone, in which very few sponges were found, water temperatures could become high enough to somewhat limit reproduction of some species. In addition, heavy matted, detached algal growths which would kill many sponges were abundant in these shallow waters.

A sand bar formed the northern boundary of this zone, but it did not, in itself, appear to limit the northerly dispersion of sponges. The southern boundary of this zone, on the other hand, constituted the northern distribution limit for a number of species (Fig. 2), and may be directly related

to limiting temperature values. Two strong offshore eddies
channelled the flow of water towards the open sea (Fig. 1),
affecting northern larval dispersion. Furthermore, the water
turbulence in the area resulted in a sediment level in the
water high enough to cover and kill some sponge species.

ZONE E: Probably as a result of the barriers to dispersion
found in the shallow water areas of Zone D, Zone E was the
northern distribution limit for a number of sponge species.
The high number of species represented in this zone (a total
of 28 species, 18 of the significantly distributed species)
was probably a reflection of a combination of factors. Rock
bars were abundant in this area, and were found in water
depths varying from less than 20 feet to more than 50 feet.
A yearly mean temperature of 73.5°F, with a lowest mean
temperature of 62.5°F was favorable for both growth and reproduction of most sponges, and nutrients were supplied by
the rivers flowing into this zone. In addition to the strong
parallel inshore current, two eddies were located.

ZONE F: The ecology of this zone appeared to differ considerably from that of Zone E, and as a result the species
represented in the two zones were quite different. The
northern boundary of Zone F constituted the northern distribution limit for three of the significantly distributed
species.
   The water surface temperatures reported for Zone F (a
yearly mean of 73.5°F with a lowest mean temperature of
62.5°F) were favorable and no different than those reported
for Zone E. However, Zone F was a shallow area. Instead,
Zone F received nutrients from the Anclote River, rich in
nutrients because of the city of Tarpon Springs. This, along
with many rocky bars, promoted excellent wool sponge growth
and created a favorable niche for several sponge species.
In this area a number of sponge species were often crowding
each other on the rocky bars. This zone was also somewhat
removed from the main inshore current flow.

ZONE G: This zone provided diverse and generally favorable
conditions for sponge growth and reproduction and many
species were found here (27 total, 24 of the significantly
distributed species). This zone established a northern
boundary for four significantly distributed species and a
southern distribution limit for several. High nutrient
levels probably resulted from the several sewage outfalls in
the area.
   The reported yearly mean temperature in Zone G was 75.6°F,

with a lowest mean temperature of 65.3°F. This range would provide optimum sponge growth and reproduction for many species. The bottom slope in this zone was relatively steeper, and sponges were found in depths varying from less than 20 feet to more than 50 feet.

ZONE H: Many species found in the more northern Zone G were not found in Zone H (a total of 15 species in Zone H compared to a total of 27 in Zone G).

Currents in this zone (Fig. 1) would probably tend to carry sponge larvae out towards the open sea, rather than towards Zone G. In general this area appeared to have large amounts of sediment carried into the area either by the tidal wash in and out of Tampa Bay or brought northward from Zone I. Since this southerly zone has few sponges and the dominant currents are the north and/or seaward, these may have been the factors in limiting sponge species.

ZONE I: Vary few sponge species were found in Zone I, although a high number were observed in the southern limits of this zone. South of Fort Meyers increasing temperature was not responsible for restricting sponge distribution in this zone.

Water in this zone was deep close to shore and very few rock bars were found. Probably the most detrimental environmental factor of this zone was the high turbulence caused by greater wave activity, which in turn resulted in a sediment level so unfavorably high as to make the water appear quite 'murky'.

SUMMARY

It has been observed by the author that the most favorable areas for finding a wide variety of species is often in locations where there is a tidal flow combined with minimum sedimentation rates, low wave activity, and moderately warm water temperatures. In addition, the area must possess many rocky bars which allow for sufficiently high concentrations of sponges to promote high reproduction rates. Much of what has been discussed above regarding the factors delineating the zones has had to be derived from direct observation rather than hard data. The most valuable contribution of such observations is in directing attention toward the kind of research which needs to be carried out in the future.

Other factors than those mentioned above are also important to sponge ecology. Nutrients play an important role. For example, off the west coast of Florida where the nutrient

waters from the Everglades enter the Gulf, individual specimens of Spheciospongia will be found almost 5 feet in height
and 3-4 feet across. In the Bahamas, where nutrients are
low, the same species varies from 2½ by 2½ feet to ones which
are 4-5 feet across but low and doughnut shaped, an indication of poor nutrient conditions. Likewise, higher temperatures are important both to rate of growth and early maturation as seen in the wool sponge.

Wave activity is important to some species. These are
usually found in such areas as the coral reef or shallow
rocky bars exposed to high tidal and wave activity.

Various combinations of these ecological factors are found
along the Gulf coast of Florida and it is the occurrence of
one or more favorable factors in any one zone which results
in greater diversity and higher concentrations.

ACKNOWLEDGMENTS

The original data used was collected while the author was
being supported on a grant from the U.S.F.W.S. in 1955-57.
Mr. Robert Work of the University of Miami, Rosenstiel
School of Marine and Atmospheric Sciences, was the assistant
and responsible for the identification of the sponges.

LITERATURE CITED

Crawshay, L.R. 1939 Studies in the market sponges. I.
  Growth from the planted cuttings. J. Mar. Biol. Assn.,
  U.K., 23: 553-574.
Moore, H.F. 1910 A practical method of sponge culture.
  Bull. U.S. Bur. Fish., 1908, 28: 545-586.
Rathbun, R. 1887 The sponge fishery and trade. Fisheries
  and Fishery Industries of the U.S., 2: 817-841.
Society of Economic Paleontologists and Mineralogists 1955
  Finding ancient shorelines. Soc. Econ. Paleo. & Mineralogy,
  Tulsa, Okla., Special Pub. #3.
Storr, J.F. 1964 Ecology of the Gulf of Mexico commercial
  sponges and its relation to the fishery. U.S. Fish &
  Wildlife Serv. Special Scientific Report - Fisheries No.
  466, Washington, D.C.
U.S. Department of Commerce 1955 Surface water temperatures
  at tide stations, Atlantic Coast, North and South America.
  U.S. Dept. Commerce, Coast and Geog. Survey, Spec. Publ.
  No. 278, 5th ed.

FIELD OBSERVATIONS OF SPONGE REACTIONS
AS RELATED TO THEIR ECOLOGY

John F. Storr

Department of Biology
State University of New York at Buffalo
Buffalo, New York 14214

ABSTRACT: A number of field observations are reported
on: withdrawal of living sponge tissue as a reaction to
lower salinity water; healing of the cut surface of
sponges in 2-3 days; and modifications of sponge shape
from the expected due to low nutrient supply. Also discussed is the observed contraction of sponges upon cutting and the possible connection to backwashing from the
incurrent ostia, a method used by Spheciospongia vesparium to clear the subdermal chambers of soft material
drawn into the sponge in the incurrent stream.

Little has been said and less done to document some basic
reactions of sponges in their natural environment. This report is one of field observations of some sponge activities
in the hope that they will promote sufficient interest to
encourage formal research into these phenomena.
   Four separate observations have been made: withdrawal of
sponge tissue from the surface to a lower depth as a result
of contact with lower salinity water; healing of cut
surfaces; contraction of sponge material; response to low
nutrient supply; and, of greatest interest, backwashing
through the incurrent pores to rid the sponge of sedimentary
materials settling out of the water during turbid water conditions.

Tissue Withdrawal
   This has been observed in the wool sponge (Hippiospongia
lachne) and was observed in the Gulf of Mexico shortly after
the passage of a hurricane which was accompanied by heavy
rainfall. In this case the living material retreated from
the surface leaving the ends of the spongin fibers exposed.
Retreat was less than one cm. In view of experiments with
healing of cut surfaces, such a retreat may take only a
matter of a day or two. This retreat is illustrated in
Figure 17, page 50 (Storr, '64).

Healing of Cut Surfaces
   A number of experiments have been carried out to observe

the process of the healing of cut surfaces, primarily in the sponge <u>Ircinia</u> <u>strobilina</u> and other massive sponges (Fig. 1)

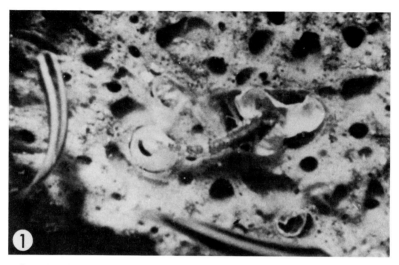

Fig. 1. Healing of cut sponge surface–newly cut surface showing internal canals. Secreted tubes of polychaete worms in some canals. Small fish feeding on commensal organisms.

In the figure a number of small fish can be observed gathered on the cut surface to pick up the many small organisms living in the canals which have been exposed by cutting. Part of a membranous tube formed by a polychaete worm is also exposed. After part of the sponge was removed, there was a contraction of the sponge material so that the more compact spongin material of the primary fibers was left raised above the originally flat cut surface. The first sign of healing was a glazing over of the cut surface in the first few hours producing a smooth shiny surface. Within 24 hours this surface had darkened but was not quite as dark as the normal surface (Fig. 2). The "skin" thus formed was thin in appearance and the cut ends of the ascending fibers were still ragged and uneven in appearance. These cut ends form the basis of the new conules while the thin skin has grown over most of the open canals leaving only a few openings unsealed. By the second day (Fig. 3) the surface was almost identical to the normal surface and by the third day the cut surface was indistinguishable from the remainder of the sponge.

On several occasions an area of the sponge was cut off several times in succession and healing appeared to take

Fig. 2. Healing of cut sponge surface after one day. Thin light grey "skin" covering most internal canal openings with new conules formed by shrinking of tissues away from the spongin fibers.

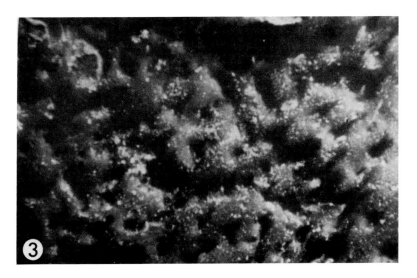

Fig. 3. Healing of cut sponge surface after 2 days. Surface almost identical to remainder of sponge.

place at the same rapid rate. While the healing of the cut surface, however, may simply be a change in cell type with little loss of energy, there could be a drain on the energy reserves of the entire sponge. It has been observed that heavy production of larvae in a sponge will use up some of the living tissue of the sponge, particularly in the central area.

Response to Low Nutrient Supply

The species observed have been Hippiospongia lachne and Spheciospongia vesparium. In both these sponges, and probably in others, the sponge will grow to a specific maximum size and at that time the central area of the sponge will begin to die off resulting in a doughnut shape. In the wool sponge, Hippiospongia lachne, the maximum size is approximately 30-35 cm in diameter and in the loggerhead, Spheciospongia vesparium, greater than one meter in diameter. Apparently the major filtering action and removal of food takes place near the outer layer of the sponge resulting in starvation in the central portion. In the loggerhead the appearance of the sponge changes radically from location to location as has also been noted by Wiedenmayer in the western Bahamas. One must assume that some of this variation is due to food supply as well as physical locality. It has been observed also that where nutrients are low and the strongest current is from one direction many loggerheads in the area are crescent shaped with the convex of the crescent facing the direction of the current and the osculi low on the down-current side. Again this growth shape would be explained on the basis of nutrient supply with a heavier supply of food being partially forced into the sponge on the up-current side promoting growth.

Contraction of Sponge Material

This is a common occurrence among many species with the amount of contraction varying from species to species. While a few sponges such as Dysidea fragilis have the ability to contract a membranous oscular chimney, the contraction in part of or all of a sponge is commonplace. A cutoff portion of a sponge will contract sufficiently to cause the cut surface to become convex in shape and the whole piece to be smaller than the area from which it was cut. The use for such contraction is difficult to establish but is perhaps associated with other functions rather than some kind of protective reaction to cutting which must be considered an unusual occurrence. Such slow contractions, for example, would be of little use in protecting a sponge from being eaten by a fish (as in the case of the loggerhead).

Backwashing
    The most unusual sponge reaction encountered was back-
washing. This has only been observed in the species Spheciospongia vesparium, the loggerhead sponge. Backwashing can be
induced by placing a handful of soft detrital material or mud
on the upper surface of a large loggerhead so that material
is drawn into the incurrent pores or ostia. Moments later
strong reverse flows from the ostia on the sides of the
sponge eject the material, similar in appearance to that of
a puff of smoke (Figs. 4-5). Whether this sudden reverse
flow is caused by current reversal or by sudden contraction
in the area could not be determined. Certainly the reaction

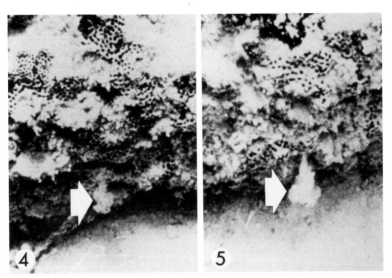

Figs. 4 & 5. Backwashing in Spheciospongia vesparium. Fig.
4 shows the start of a "puff" of backwashed material from an
incurrent pore. Fig. 5 captures a backwashing at its maximum. The strength behind the backwash is sending the material 7-10 cm out from the sponge. Arrows designate backwashed
material in each figure.

is involved in cleansing, which in some other sponges is
performed by secretions carrying the settled material out of
the ostia (e.g. Condrilla nucula).

ACKNOWLEDGMENTS

I have used the taxonomic revision of the Bahamian sponges by Felix Wiedenmayer for reference and taxonomic nomenclature. James Stamos, illustrator for the Biology Division, reproduced the illustrations from my 16 mm movie film footage.

LITERATURE CITED

Storr, John F. 1964 Ecology of the Gulf of Mexico. Commercial Sponges and its Relation to the Fishery. Special Scientific Report-Fisheries No. 466, 73 pp., U.S. Dept. Interior, Fish & Wildlife Service.
Wiedenmayer, Felix 1976 Monograph of the Shallow Water Sponges of the Western Bahamas (unpublished manuscript).

SPONGE FEEDING: A REVIEW WITH A
DISCUSSION OF SOME CONTINUING RESEARCH

Thomas M. Frost

Department of Biological Sciences
Dartmouth College, Hanover, New Hampshire 03755 USA

ABSTRACT: A brief review of the physiology of demosponge feeding is provided. Published investigations of
the rates of water transport are reviewed and their results are presented as a normalized rate of ml of water
filtered per second per ml of sponge. Values for these
rates ranged from 0.002 to 0.84. In situ measurements
of filtering rates by Spongilla lacustris are described.
Values between 0.027 and 0.009 ml/sec/ml were recorded
for 6 sponges in October 1974. Some continuing ecological research on freshwater sponge feeding is briefly
discussed.

INTRODUCTION

Despite the prevalent view of sponges as "simple" organisms, the Porifera exhibit intricate feeding mechanisms.
Effectively, a sponge presents a network of progressively
smaller filters to the flow of water generated through its
canal systems. Investigations of the magnitude of this flow
of water are the primary focus of this paper. I have attempted to describe all such measurements as have been reported
in the literature.

By way of introduction, however, a brief discussion of
the physiology of feeding is appropriate. I have not attempted to review all of the available literature in this area
but the most current and salient papers regarding these processes are briefly described. Also, this discussion will be
confined to the Demospongiae which comprise at least 95% of
the species of sponges (Reiswig, '75a) and on which the
great majority of work has been focused.

PHYSIOLOGY OF SPONGE FEEDING

Van Trigt ('69), van Weel ('49), Kilian ('52), Schmidt
('70) and Reiswig ('71b, 75a) have provided descriptions of
the overall feeding systems of demosponges and the following

outline is a consensus of their observations.

A current of water is developed by the activity of the flagella of the choanocytes and this flows through the aquiferous system along a gradient of decreasing resistance to flow. The hydromechanics of this system have been analysed by Bidder ('23, '37), Leigh ('71), Vogel ('74) and Reiswig ('75a). Reiswig ('75a) calculated that laminar flow occurred throughout the canal system. The points of entrance for this current are the ostia on the epithelium of the sponge. These open into a broad subdermal space. The current then flows into inhalent openings on the floor of the subdermal cavities and continues through the incurrent canal system, which spreads into many progressively narrower branches eventually leading to the prosopyles of the flagellated chambers. These chambers, lined by choanocytes, mark the end of the incurrent system. The large number of these chambers occurring in the body of the sponge (e.g. about 10^7 chambers/ml in <u>Halichondria panicea</u>, <u>Haliclona permollis</u> and <u>Microciona prolifera</u>, Reiswig ('75a) suggests the fineness of the divisions which occur in the incurrent system.

The pinacocytes which make up the epithelium and line the incurrent canals and the choanocytes of the flagellated chambers are capable of phagocytosis and function as filters which remove material from the incurrent flow. Reiswig ('71b) characterized the minimum diameter of material removed by the components of the feeding system of three tropical marine sponges as greater than 50 μm for the ostia, between 2-5 and 50 μm for the incurrent canal and 0.1 μm for the collars of the choanocytes.

Material which has been taken up is transferred by a direct exchange to amoebocytes which then provide nutritive substances to other specialized cells throughout the sponge. Digestion appears to occur wholly intracellularly (Schmidt, '70). Jones ('62) and Pavans de Ceccaty ('74) have reviewed the types of interchanges which occur between cells among the Porifera, and they have included detailed discussions of the exchange of nutrient material. Many of the specialized sponge cell types have been described in detail by Simpson ('68). The relationship between the types of particle ingested, the retention time of this material and the form of the sponge feeding system has been described for <u>Mycale</u> sp., <u>Verongia gigantea</u> and <u>Tethya crypta</u> (Reiswig, '71b).

Removal of material from a sponge occurs by a reversal of the particle uptake processes. Undigested substances are transferred by amoebocytes to the pinacocytes lining the excurrent canals and then released to the excurrent stream by a form of reverse phagocytosis (van Weel, '49). The excurrent stream is a continuation of the feeding current which has exited from the apopyle of the flagellated chamber. The excurrent portion of the aquiferous system is similar to the canal system of the incurrent portion. Leading from the flagellated chambers, fine canals join into progressively larger ducts which eventually anastamose into an atrial cavity with a specialized opening to the outside, the osculum. The volume of the components of the canal system reaches its peak at the flagellated chambers. Proceeding further there is a reduction in the cross sectional area of the entire excurrent system reaching a minimum at the osculum, thus facilitating a high flow rate at this point which may be important in insuring the removal of previously cycled materials from the area of supply (Bidder, '23). Reiswig ('75a) provides detailed, quantitative descriptions of the components of the feeding system in three marine sponges.

A number of investigators have focused their attention on the role of the pinacocytes and the epithelium in sponge feeding. Harrison ('72b) described two types of ostial openings in the epithelium of the freshwater sponge Corvomeyenia carolinensis: one between pinacocytes and the other within specialized pinacocytes, the porocytes. He also observed that marginal cells of the basal pinacoderm of this sponge actively took up bacteria by filopodial extension and subsequent phagocytosis (Harrison, '72a). Simpson ('63) observed similar bacterial uptake in the marine species Microciona prolifera. In the same species Bagby ('70) observed that the main specialization of pinacocytes was a coat of material on the surface which was implicated as an adhesive which held particles prior to phagocytosis.

The phagocytic ability of pinacocytes is important in preventing clogging of the sponge feeding system (Schmidt, '70, Reiswig, '71b) and may be important in maintaining the aufwuchs-free epithelium common to many freshwater sponges (van Weel, '49, Frost, '76).

In Ephydatia fluviatilis and E. mulleri, Schmidt ('69) observed that baker's yeast taken up by pinacocytes was quickly removed from these cells and that no digestion of this material was apparent. She concluded that the nutritional value of this phagocytosis was minimal and that its primary role was clearing the feeding system. In contrast, Kilian ('52) stated that phagocytic uptake by pinacocytes was the only source of nutrition for freshly germinated sponges of E. mulleri. Harrison ('72a) described a high acid phosphatase activity in pinacocytes of C. carolinensis indicating the potential for digestion of ingested material. In a budget of the material retained by three marine sponges,

Reiswig ('71b) found that approximately 80% of the particulate material utilized was of a size which was likely to be taken up mainly by the choanocytes. For Haliclona permolis and Suberites fiscus in organic-rich estuaries he found that the entire nutritional requirements of these sponges could be satisfied by bacterial uptake ('75b) which would occur primarily at the choanocytes. Phagocytic activity by pinacocytes on the epithelium and within the aquiferous system of sponges is important although its contribution to nutrition is not fully understood.

The choanocytes are certainly the keystone cells of the sponge feeding system. The movement of their flagella drives the water current through the system and the fibrils of their collars provide a tremendous surface area for filter feeding. Reiswig ('75a) has calculated that the intervillar space provided by these fibrils in three marine demosponges is 12 to 56 times larger than their total external surface area. Electronmicrographic studies by Rasmont et al. ('58), Rasmont ('59) and Fjerdingstad ('61) have revealed that the collars of these choanocytes consist of a series of fibrillar projections spaced approximately 0.1 μm apart. Fjerdingstad also noted that these projections were interconnected by microfibrils.

Uptake at these fibrillar projections has been observed by many investigators. Van Trigt ('19) observed the uptake of carmine particles and zoochlorellae which had previously been isolated from a sponge by the choanocytes of Ephydatia fluviatilis and Spongilla lacustris. van Weel ('49) used a variety of materials: carmine, India ink, blood, crushed testes of toads, milk and a 0.1% solution of iron sugar to view uptake on the collar cells of Spongilla proliferans. Similarly, Kilian ('52) observed the uptake of carmine, India ink, bacteria, algae, milk, and albumin by E. fluviatilis. Using this same species and E. mulleri, Schmidt ('69) examined the accumulation of baker's yeast ·and fluorescently labelled bacteria and casein.

The movement of the flagellum of a choanocyte has been described by Kilian ('52) and Rasmont ('59) as occurring in one plane although there was no clear pattern to the movement of flagella within a chamber. Feige ('69) described appendages on the flagella of Ephydatia fluviatilis and proposed that these served to increase its efficiency in generating a water current.

WATER TRANSPORT RATES

Examinations of water transport rates can be grouped into

two categories. Most studies have involved estimation of the volume of water escaping from an osculum. Others have derived their estimates by measuring the effect of a sponge on the concentration of materials present in a chamber containing a known volume of water. Usually the latter investigations yield a clearance rate which can only be roughly converted to an absolute rate of water transport by assuming a value for the efficiency of retention for measured materials.

I have described the investigations of water transport rates which have been reported in the literature. The values calculated by these authors are listed as reported and, to facilitate comparison, as a standard rate in terms of ml of water filtered per second per ml of sponge (ml/sec/ml). These converted values are in Table 1. The species names are those used by the authors and in some cases do not conform to the current taxonomy.

Studies in which there was a measurement of flow are described first, and the temperature at which the rate was measured is provided if it was available. Pütter ('08) reports a water transport rate of 0.5 liters per hour for a 60 ml specimen of Suberites domuncula. He derived this value by estimating, by an undescribed method, a 5 mm/sec velocity for water escaping an osculum 6 mm in diameter. The transport rate reported is a maximum for he observed variations in the flow rate and, occasionally, total closure of the osculum. Converting to ml/sec/ml yields a rate of 0.002.

Parker ('14) measured the current produced by Spinosella sororia at 18°. In a laboratory he tied a glass tube into one osculum on a finger of a freshly collected colony of this sponge. He then calculated the rate of water flow out of the tube by noting the movement of suspended particles within it. The finger measured 10 cm in length and averaged 4 cm in diameter, yielding an approximate volume of 125 ml, and discharged 4.5 ml in 5 sec. This converts to 0.007 ml/sec/ml.

Bidder ('23) calculated values for water transport by a specimen of Leucandra aspera by measuring the distance traveled by water leaving its osculum. This length was traced using a litmus or carmine solution. The path length was converted to a delivery per second rate of 26 ml by a formula which he derived. The only datum available regarding the volume of the individual sponge measured is a value of 0.31 ml for the preserved specimen. Using this value a water transport rate of 0.84 ml/sec/ml can be calculated although this may be much higher than the actual value for a rate per volume of living sponge.

For a specimen of Ephydatia fluviatilis which had been

287

hatched from a gemmule 7 days before measurement, Kilian ('52) obtained a transport rate of 4.5 ml in 24 hours. The volume of water exiting the osculum was determined by timing the movement of particles in the excurrent stream from an oscular chimney with a known volume. He indirectly reported the volume of this specimen estimating that it would have a total of 1420 flagellated chambers and stating that they normally occurred at a concentration of 7600 chambers per mm^3. A volume of 0.00019 ml can be calculated from these data yielding a rate of 0.28 ml/sec/ml.

For a similar investigation Arndt ('30) reports a personal communication from Schröeder in which the water current was measured from an eight-day-old unidentified freshwater sponge which had been hatched from a single gemmule. No description of the experimental method was provided, but, assuming that this sponge had approximately the same volume as the specimen investigated by Killian (0.0002 ml), the transport rate he derived of 6.307 ml in 24 hours can be roughly converted to 0.4 ml/sec/ml.

Vogel ('74) raised an interesting question regarding the effect of currents in the environment on the water transport rate of sponges. His work indicated an increase of water transport concommitant with an increase of water flowing over the sponge. This phenomenon was attributed to a passive flow set up through the sponge due to the hydromechanical gradient in the canal system. For Halichondria bowerbanki the velocity of water exiting oscula was recorded using a thermistor flowmeter. He reported water transport rates of 2.8 to 6.8 cm/sec in external currents of 0 and 10 cm/sec, respectively. The seven sponges utilized in these studies had an average internal oscular diameter of 4.9 mm and, assuming a cylindrical shape, an average volume of 3.5 ml. A value of 0.24 ml/sec/ml can be calculated for the average of the flow rates recorded in the experiments, 4.37 cm/sec.

Reiswig ('75a) examined the water transport rate of a specimen of Haliclona permolis. His technique combined the use of photographs taken at slow speed to delineate the plume of water exiting an osculum with tubes having approximately the same diameter as the sponge osculum. A plume was developed from a tube to correspond in form to that exiting the osculum and the rate of particles moving within that tube was measured. This method yielded a rate of 0.314 ml/sec which converts to 0.132 ml/sec/ml for the specimen investigated which had a volume of 2.386 ml.

Brauer ('75) derived transport rates indirectly by measuring the rate of efflux of water from Spongilla lacustris. Young, laboratory reared sponges, which had been hatched

from gemmules five days before the experiments, were maintained in a medium containing tritiated water for 1 - 2 or
18 hours at 21 to 23°. Specimens were then transferred to
non-radioactive medium where the increase in radioactivity
was measured for 1 hour. Controls indicated that uptake of
tritiated water by cells was not important. Water efflux
rates measured in this manner indicated that the sponges
turned over the volume of water present within them 70 times
per hour on the average (n = 63). A value of between 60 to
70% was measured for the volume of the canal systems of these
sponges. Combining these two facts a water transport rate of
approximately 0.0118 ($\overline{X}$, n = 63) ml/sec/ml is estimated. Although this investigation did not involve a direct measure of
water current the values reported provide close approxima-
tions of an absolute water transport rates since assumptions
regarding efficiency of retention were not important.

The next two authors have measured water transport indirectly and provide filtering rates rather than absolute water
transport rates. Pütter ('14) determined a filtering current
for Suberites massa by measuring the removal of oxygen from
a chamber in which a sponge was metabolizing. He observed
that 1 ml of a sponge tissue weighed 1 gm and was comprised
on the average of 100 cm^2 of surface area of flagellated
chambers and canal system. The maximum value for removal of
oxygen recorded for this species was 14.5 mg O_2/m^2-hr at 24.7.
For seawater with a concentration of 7 mg/liter of O_2 a
sponge weighing 100 gm would have to pump approximately 2
liters/hr in order to make such an amount of O_2 available.
This value is an absolute minimum since it requires 100%
efficiency in the uptake and retention of O_2. Assuming such
an efficiency a water transport rate of 0.01 ml/sec/ml is
obtained.

Jørgensen ('43) reports clearance rates for three marine
sponges in terms of ml/hr/mgm amino nitrogen. His method
involved measuring the removal of carbon particles, ranging
in size from 2 to 4.5 μm, from a chamber using a photometer.
No more detailed information is provided for the experimental method he utilized. For Grantia compressa an average
rate of 180 ml/hr/mgm amino nitrogen is reported for 9 sponges and another individual exhibited a rate of 135. Values
for Sycon coronatum were between 200 and 145 for 9 specimens.
One specimen of Halichondria panicea filtered 65 ml/hr/mgm
amino nitrogen. No volumes were reported for the sponges
used in these experiments but the average length for all but
H. panicea is provided. An average diameter of 1.5 mm and a
cylindrical form have been assumed to approximate volumes
for these sponges: 12.37 and 8.84 ($\overline{X}$, n = 9) for G. com-

pressa and 0.0054 ml/sec/ml ($\bar{X}$, n = 9) for S. coronatum; these rates are absolute water transport rates if 100% retention efficiency is assumed for carbon particles.

All of these investigations involve measurements which have been obtained in a laboratory situation. I would expect that this would alter the flow rate produced. The magnitude of this change would vary depending on laboratory conditions, the length of storage prior to measurement and the nature of the sponge itself.

Calculations of the flow rate of water moving from an osculum should consider the gradient of water velocity which occurs between the center of the opening and its edge. For Haliclona permolis Reiswig('75a) determined that the effective flow across the oscular opening was 58% of the axial velocity. This problem was not considered in calculating the other flow rates measured in this fashion and therefore the values reported may be considerably higher than the actual water transport rate.

Calculating an absolute water transport rate from a filtering rate involves an estimation of the efficiency of material retention. Reiswig ('71b) recorded efficiencies of particle retention for three tropical marine demosponges and found that they varied considerably with the size and nature of the particles, averaging between approximately 60 and 98%, although overall particle retention was high, 82% by volume. Colloidal material was taken up at a much lower efficiency, averaging 35.0% with a 95% confidence interval of 29.4 to 40.8% (n = 8), even though this material comprised 80.5%, on the average, of the carbon taken up by these organisms. In another study, retention of bacteria as determined by plate counts by Haliclona permolis and Suberites fiscus ranged between 55 and 91%, averaging 77% (Reiswig '75b).

Large amounts of nutritive material may be quickly ejected without digestion by the sponges as was observed by Schmidt ('70) with baker's yeast in Ephydatia fluviatilis and E. mulleri. Similarly non-nutritive material such as carmine or carbon particles may be quickly cycled through the sponge after uptake (Van Trigt '19, van Weel '49). Therefore in an experiment with a duration longer than the turnover time of the sponge feeding system a lowered uptake and consequently a lower feeding rate will be manifested due to the rerelease of captured materials.

Reiswig ('71a) circumvented these difficulties by directly determining pumping rates for three sponges on the coral reefs off the coast of Jamaica. Large oscular openings characterize the three species which he investigated facilitating direct measurement of the excurrent flow across the en-

Table 1. Reported values for water transport rates of sponges converted to units of ml of water filtered per second per ml of sponge. The assumptions necessary for making conversions to this form are described in the text.

INVESTIGATOR	SPECIES	VOLUME OF SPONGE ml	WATER TRANSPORT RATE ml of water transported per sec per ml of sponge
Pütter ('08)	Suberites domuncula	60	0.002
Parker ('14)	Spinosella sororia	125	0.007
Pütter ('14)	Suberites massa	100	0.011[1]
Bidder ('23)	Leucandra aspera	0.31	0.84
Schröder (Arndt '30)	unidentified freshwater sponge	---	0.4
Jørgensen ('49)	Grantia compressa	12.37	0.003[1]
	Grantia compressa	8.84 ($\bar{X}$, N = 9)	0.0057[1]
	Sycon coronatum	8.84 (X, N = 9)	0.0054[1]
Kilian ('52)	Ephydatia fluviatilis	0.00019	0.28
Vogel ('74)	Halichondria bowerbanki	3.5	0.24[2]
Reiswig ('74)	Mycale sp.	---	0.24[2]
	Verongia gigantea	---	0.077[2]
	Tethya crypta	---	0.125[2]
Reiswig ('75)	Haliclona permolis	2.4	0.131
Brauer ('75)	Spongilla lacustris	---	0.0118

[1] Derived from filtering rates.
[2] Size weighted averages for natural sponge populations.

291

tire osculum. To derive these rates recording thermistor
flowmeters and hand held anemometer tube flowmeters operated
by SCUBA divers were used. Size weighted average transport
rates during the month of October in 1969 were 0.24 ml/sec/ml
for Mycale sp., 0.077 for Verongia gigantea, and 0.125 for
Tethya crypta (Reiswig '74). In conjunction with this study
Reiswig ('71b) also investigated, quantitatively and qualitatively, the particulate matter utilized by these sponges. He
also directly measured the removal of bacteria from estuarine
waters by Haliclona permolis and Suberites fiscus ('75b).
The size and morphology of these sponges permitted direct
sampling of water from incurrent and excurrent regions. The
results of these investigations provide the best information
available regarding the food utilized by the Porifera in a
natural system.

Considering the range of values which have been reported,
from 0.002 to 0.84 ml/sec/ml, it is difficult to draw general
conclusions about the feeding rates of the phylum Porifera.
This broad range is probably a reflection of the diversity of
rates occurring within natural systems and of the variety of
techniques used in measurement. Reiswig ('74) has reported
large variations for these rates in natural populations (e.g.
from 0 to 0.22 ml/sec/ml for Tethya crypta). The results of
Reiswig's in situ studies ('71a, '74) provide the most re-
liable values for water transport rates which are available.

CONTINUING RESEARCH

In order to broaden the available information on the ecology of sponge feeding, I have begun investigating feeding
activities in the freshwater sponge Spongilla lacustris. The
population of sponges being investigated occurs in Mud Pond,
a sphagnum bog-pond near West Canann, New Hampshire, U.S.A.
($43^{\circ}38'$N, $72^{\circ}05'$E). S. lacustris grows abundantly throughout
all but the deepest portions of this pond during the summer.
The seasonal occurrence and life cycle of this sponge population have been described by Simpson and Gilbert ('73, '74).
Algal symbionts produce a bright green color in this species
which typically occurs in elongate branches 1 to 8 mm in
diameter attached to macrophytes, or growing up from the pond
bottom. Many small oscula are usually distributed across the
surface of these branches. This growth form precludes any
direct in situ measurements of the rate of water transport.

Estimates of filtering rates have been obtained by indirectly measuring the uptake of radioactively labelled bacteria
by sponges of known volume. This method involves measuring
the change in the concentration of ^{32}P labelled Aerobacter

in a chamber in which a sponge has been filtering.

The uptake of particles by the sponge effects an exponential decay in the concentration of these particles in the feeding chamber. Such a decay is described by the general equation.

$$C_i = C_o e^{-kt}$$

where C_i = the concentration after i units of time

C_o = the concentration at time 0

k = the relative removal rate per t

t = i units of time

Manipulating this equation algebraically, one obtains an equation of the form y = mx + b where k = m, the slope of the line:

$$\ln C_i = -kt + \ln C_o$$

An equation of this form lends itself to linear regression analysis: regressing the ln of the values obtained for the concentration of particles with time yeilds k as the slope of the regression line. The relative rate of removal may be converted to a filtering rate/unit volume by multiplying by the volume of water in which the sponge was feeding and dividing by the volume of the sponge. This filtering rate may be converted to a water transport rate by dividing by the efficiency of particle uptake.

Preliminary experiments utilizing these techniques were performed at Mud pond on 1 and 17 October 1974. Water temperatures on these dates were $13.2^{\circ}C$ and $9.2^{\circ}C$ respectively. Branches of <u>Spongilla</u> <u>lacustris</u> were gathered from the pond and suspended in B.O.D. bottles filled with 250 ml of pond water. These bottles were floated in the pond on a raft and the sponges were permitted to acclimatize for longer than 15 minutes. After this period ^{32}P labelled bacteria were added producing a concentration of approximately 3 x 10^4 cells/ml. An equal number of control bottles without sponges were also innoculated. Samples were taken from these chambers every 15 min for 2 hours. One 0.3 ml sample was taken at each time point on 1 October, three 0.1 ml samples were taken at each time point on 7 October. Radioactivity per sample was measured on a Nuclear Chicago End Window Counter after either filtering the sample onto a 0.2 μm pore size Millipore filter or drying a portion of it on a planchet. On both dates feeding chambers containing sponges exhibited an exponential decay in bacterial concentration but in no instance was there a significant decrease in the concentration in the control bottles. Filtering rates were calculated by regressing the radioactivity per sample on the time of sampling. The results of these calculations are listed in Table 2.

Table 2. Filtering rates of Spongilla lacustris.

SPONGE VOLUME ml	FILTERING RATE ml/sec/ml
	1 October
1.6	0.027
2.8	0.023
3.0	0.011
	17 October
8.6	0.010
9.8	0.010
7.9	0.009

Assuming an 80% efficiency in bacterial retention these filtering rates range between 0.034 and 0.011 ml/sec/ml, falling within the range of values reported in Table 1. They conform very well with the rate obtained by Brauer ('75) for S. lacustris. However, her experimental conditions, utilizing a laboratory germinated 5 day old sponge, would not necessarily simulate a natural situation.

In a comparison to the only other in situ study, these rates are substantially lower than those obtained by Reiswig ('71a) for three, large tropical marine demosponges. The difference in temperature at the time of measurements may account for these differences. Reiswig's values are for a temperature of $29.5^\circ C$. Possible sources of error in my values include the estimate of filtering efficiency and bottle effects. However, it seems likely that S. lacustris would not exhibit a water transport capacity similar to these tropical species. Either the great difference in size or the differences in habitat may explain the lack of similarity. Furthermore, Gilbert and Allen ('73a, '73b) have found that the symbiotic algae within S. lacustris contribute photosynthate to it and that during daylight periods

this association may exhibit a net primary production, possibly reducing a need for feeding activities.
These data represent the results of preliminary experiments and more detailed investigations are now in progress. An apparatus has been developed which allows a constant gentle stirring of the feeding chamber in situ. This provides a more homogeneous environment for feeding experiments, hopefully lessening bottle effects, and permits the use of larger particles such as algae and yeast without settling. Scintillation counting techniques have been incorporated permitting more accurate measurements of concentration and the simultaneous measurement of two types of particles by double isotope counting techniques.

Using these improved methods a number of questions can be asked. What is the importance of particle size and particle concentration on filtering rate? What are the effects of temperature or life cycle stage? Are there any daily patterns in the filtering activities?

The zoochlorellae present within Spongilla lacustris may have an important effect on the filtering activity of this sponge. The nature of the sponge-algal symbiosis affords an opportunity to assess the contribution of the algae to the association. Viable aposymbiotic sponges have been grown and maintained in situ through the use of a raft which shields the sponges from sunlight but permits nearly normal interchange of water with the pond (Frost, in preparation). Growth rates and data for chlorophyll and pheopigment content of these sponges and specimens grown under comparable conditions in light have been obtained. Experiments comparing the feeding rates of normal and aposymbiotic sponges are in progress. These will be coupled with primary productivity studies to assess the nutritional role of each member of the association.

Finally, the natural occurrence of particles in Mud Pond will be evaluated in order to develop a nutritional budget for Spongilla lacustris in this system.

ACKNOWLEDGMENTS

I would like to express my gratitude to John Gilbert, Michel Kabay, James Litton and Peter Starkweather for their assistance in the preparation of this manuscript.

LITERATURE CITED

Arndt, W. 1930 Schwämme. pp. 39-120 In: Tabulae Biologicae W. Junk (Ed.), Supplement 11 (Bank VI). W. Junk,

Berlin 969 pp.

Bagby, R. M. 1970 The fine structure of pinacocytes in the marine sponge Microciona prolifera (Ellis and Solander). Z. Zellforsch., 105:579-594.

Bidder, G. P. 1923 The relation of the form of a sponge to its currents. Quart. J. of Micros. Soc., 67:293-325.

_____1937 The perfection of sponges. Proc. Linnean Soc. London, 149:119-146.

Brauer, E. B. 1975 Osmoregulation in the fresh water sponge, Spongilla lacustris. J. Exp. Zool., 192:181-192.

Feige, W. 1969 Die Feinstruktur der Epithelien von Ephydatia fluviatilis. Zool. Jb. Anat., 86:177-237.

Fjerdingstad, J. 1961 The ultrastructure of choanocyte collars in Spongilla lacustris. Z. Zellforsch., 53:645-657.

Frost, T. M. 1976 Investigations of the aufwuchs of freshwater sponges. I. A comparison between Spongilla lacustris and three aquatic macrophytes. Hydrobiologia, in press.

Gilbert, J. J., and J. L. Allen. 1973a Studies on the freshwater sponge Spongilla lacustris. Organic matter, pigments and primary productivity. Verh. Internat. Verein. Limnol., 18:1413-1420.

_____1973b Chlorophyll and primary productivity of some green, freshwater sponges. Int. Revue ges. Hydrobiol., 58:633-658.

Harrison, F. W. 1972a The nature and role of the basal pinacoderm of Corvomeyenia carolinensis Harrison (Porifera: Spongillidae). A histochemical and developmental study. Hydrobiologia, 39:495-508.

_____1972b Phase contrast photomicrography of cellular behaviour in spongillid porocytes (Porifera: Spongillidae). Hydrobiologia, 40:513-517.

Jones, W. C. 1962 Is there a nervous system in sponges? Biol. Rev., 37:1-50.

Jorgensen, C. B. 1949 Feeding rates of sponges, lamellibranchs, and ascidians. Nature, 163:912.

Kilian, E. F. 1952 Wasserströmung und Nahrungsaufnahme beim Susswasserschwamm Ephydatia fluviatilis. Z. vergl. Physiol., 34:407-447.

Leigh, E. G. 1971 Adaptation and Diversity. Freeman, Cooper & Company. San Francisco, California. 288 pp.

Parker, G. H. 1914 On the strength and the volume of the water currents produced by Sponges. J. Exp. Zool., 16:443-446.

Pavans de Ceccatty, M. 1974 Coordination in Sponges. The foundation of integration. Amer. Zool., 14:895-903.

Pütter, A. E. 1908 Studien zur vergleichenden Physiologie des Stoffwechsels. Abh. Ges. Ws. Gottingen Math-phys Kl., N. F., 6:1-79.
_____1914 Der stoffwechsel der Kieselschwamme. Z. allg. Physiol., 16:65-114.
Rasmont, R. 1959 L'Ultrastructure des choanocytes d'éspon- ges. Ann. Sci. Nat. Zool., 12:253-262.
Rasmont, R., J. Bouillon, O. Castiaux, and G. Vandermeersche. 1958 Ultrastructure of the choanocyte collar-cells in freshwater sponges. Nature, 181:58-59.
Reiswig, H. M. 1971a In situ pumping activities of trop- ical Demospongiae. Mar. Biol., 9:38-50.
_____1971b Particle feeding in natural populations of three marine Demospongiae. Biol. Bull., 141:568-591.
_____1974 Water transport, respiration and energetics of three tropical marine sponges. J. Exp. Mar. Biol. Ecol., 14:231-249.
_____1975a The aquiferous systems of three marine demo- spongiae. J. Morph., 145:493-502.
_____1975b Bacteria as food for temperate-water marine sponges. Can. J. Zool., 53:582-589.
Schmidt, I. 1970 Phagocytose et pinocytose chez les Spongillidae. Z. vergl. Physiologie., 66:398-420.
Simpson, T. L. 1963 The biology of the marine sponge Mic- rociona prolifera (Ellis and Solander). I. A study of cellular function and differentiation. J. exp. Zool., 154:135-152.
_____1968 The structure and function of sponge cells: New criteria for the Taxonomy of Poecilosclerid sponges. Pea- body Mus. Natur. Hist. Yale Univ. Bull., 25:1-141.
Simpson, T. L., and J. J. Gilbert. 1973 Gemmulation, gem- mule hatching, and sexual reproduction in freshwater sponges. I. The life cycle of Spongilla lacustris and Tubella pennsylvanica. Trans. Amer. Micros. Soc., 92: 422-433.
_____1974 Gemmulation, gemmule hatching and sexual repro- duction in freshwater sponges. II. Life cycle events in young, larva-produced sponges of Spongilla lacustris and an unidentified species. Trans. Amer. Micros. Soc., 93:39-45.
Vogel, S. 1974 Current-induced flow through the sponge Halichondria bowerbanki. Biol. Bull., 147:443-456.
Van Trigt, H. 1919 A contribution to the physiology of the freshwater sponges (Spongillidae). Tijdschr. Ned. Dierk. Ver., 17:1-220.

Weel, P. B. van. 1949 On the physiology of the tropical freshwater sponge, _Spongilla proliferans_ Annandale. I. Ingestion, digestion and excretion. Physiol. Comp., 1:110-128.

LIFE CYCLES OF INVERTEBRATE PREDATORS
OF FRESHWATER SPONGE

Vincent H. Resh[1]

Department of Biology, Ball State University
Muncie, Indiana 47306

ABSTRACT: Life cycles of several aquatic insects are
linked to the bionomics of their food sources, freshwater
sponge (Porifera: Spongillidae). Caddisflies (Trichoptera: Leptoceridae) that feed on sponges have five larval instars but divergent life cycles. Ceraclea resurgens (Walker) has a univoltine pattern with a brief
spring emergence period. Ceraclea transversa (Hagen) has
two distinct cohorts, the first of which resembles a univoltine life cycle. The individuals of the first cohort
feed on freshwater sponge, overwinter as prepupae, and
emerge the following spring. Larvae of the second cohort feed on sponge through summer and autumn, but overwinter as detritus feeders when the sponge gemmulates.
Pupation and emergence occur the following summer. Spongilla fly (Neuroptera: Sisyridae) genera, Sisyra and
Climacia, have three larval instars, a terrestrial pupa,
and populations composed of multiple cohorts. Sponge-
feeding midges (Diptera: Chironomidae), Demeijerea rufipes (L.) and Xenochironomus xenolabris (Keiffer) have
four larval instars and adult emergence and oviposition
occur throughout the summer. Life cycles of sponge feeding populations may vary according to the gemmulation-
proliferation pattern of the host sponge in different localities. Host specificity of sponge-feeding invertebrates may be related to olfactory identification. Future studies should emphasize the morphological and metabolic mechanisms of the sponge feeding habit.

Benthic macroinvertebrates have usually been classified into broad categories in regard to their feeding habits. Genera, families, and occasionally even orders of aquatic insects have been generalized as being carnivores, herbivores,
or omnivores. There appears however, to be an inverse relationship between the size of the taxonomic group for which

[1] Present Address: Division of Entomology and Parasitology
University of California, Berkeley, California 94720

299

the generalizations are made, and the number of individual species in those groups for which the feeding habits are known. That is, among invertebrate genera and families in which the food habits of individual species have been studied, generalizations where feeding preferences can be applied to all species are rare.

One of the most unusual food sources available to aquatic macroinvertebrates is freshwater sponge. There has been some speculation that sponges are one of the only groups of animals that has evolved defensive mechanisms enabling them to be free of predators. Support for this hypothesis has been based on a variety of factors ranging from the mechanical difficulties of spicule digestion to the presence of metabolic poisons within the sponge. In actuality, this hypothesis concerning freedom from predators is not true for either freshwater sponges (Family Spongillidae) or any of the marine sponges.

The ability to feed on freshwater sponge is exhibited by certain species in single families of three different insect orders, the Trichoptera, Neuroptera, and Diptera. However, attempts to apply generalizations of sponge feeding ability to the generic level of these families would not be valid, since in two of these families, genera that contain sponge feeding species also contain non-sponge feeding species. This study deals with the life cycles of individual species that have evolved the ability to feed on freshwater sponge. However, since the life cycles of these species are so closely linked to the biology of the sponge, application of these patterns to systematically related non-sponge feeding species cannot be made.

In addition to predator-prey relationships, benthic macroinvertebrates may develop an association with freshwater sponge in two other ways. In the first case, the sponge serves merely as substrate for attachment and this type of association is typical for a great many species of aquatic insects such as mayfly nymphs, hydropsychid caddisfly larvae, and midge larvae (Roback '68). A second type of non-predatory association involves diverse organisms such as protozoans, nematodes, oligochaetes, and water mites which use the vascular system and irregular surfaces of the sponge as suitable permanent habitats (Steffen '67). While further study may indicate that some of these associations may be specific to sponges, this study will concentrate on the life cycles and habits of the caddisflies, spongilla flies, and midges that have evolved definite and established feeding associations with freshwater sponges.

ORDER TRICHOPTERA
FAMILY LEPTOCERIDAE
GENUS CERACLEA

References to caddisflies found in association with fresh-water sponge have been mainly to species in the family Leptoceridae (Siltala '07; Ross '44; Nielsen '48; Tsuda and Kuwayama '54; Satija '64; Lepneva '64; Maitland '66, '67; Roback '68; Lehmkuhl '70; Resh and Unizicker '75), although the earliest North American reference by Krecker ('20) was misidentified as being a species in the family Rhyacophilidae when in actuality it was also a leptocerid. Atypical occurrences of caddisflies in freshwater sponge have also been reported for the caddisfly families Polycentropodidae (Hicken '67) and Limnephilidae (Clady '69).

Table 1. Distribution of sponge feeding species of the caddisfly genus Ceraclea based on records of Morse (1975) and those of the author.

Species	Distribution
Fulva Group:	
Ceraclea biwaensis (Tsuda and Kuwayama)	Japan
C. alces (Ross)	North America
C. resurgens (Walker)	North America
C. transversa (Hagen)	North America
C. fulva (Rambur)	Western Palearctic Region
C. albimacula (Rambur)	Western Palearctic Region
C. cama (Flint)	North America
C. vertreesi (Denning)	North America
C. latahensis (Smith)	North America
C. alboguttata (Hagen)	Western Palearctic Region
Senilis Group:	
C. senilis (Burmeister)	Western Palearctic Region
C. spongillovorax (Resh)	North America
Nigronervosa Group:	
C. nigonervosa (Retzuis)	Transcontinental Palearctic Region, North America

In the family Leptroceridae, sponge feeding appears to be limited to the monophyletic genus Ceraclea. The ancestor of this genus was probably the first caddisfly species to have evolved the ability to ingest whole particles of freshwater sponge. While representative species in all three subgenera of Ceraclea (Ceraclea, Athripsodina, and Pseudoleptocerus) have the ability to use sponge, only certain descendents of the original pre-Ceraclea ancestor have retained the sponge feeding habit in their life cycles. These species are listed in Table 1.

All members of the Fulva Group (Table 1) in the subgenus Ceraclea whose food habits are known, are obligatory sponge feeders. Morphologically, this group has evolved several modifications in the larval stage that distinguish it from other Ceraclea species. The larvae lack the characteristic parafrontal sutures that are usually described as typical of Ceraclea (= Athripsodes, in part) in most generic keys. A particularly unusual development has been the reduction in the antennal length of the last instar larvae, since long antennae are characteristic of leptocerid caddisflies. In addition, the larvae of this group construct cases made entirely of a silk secretion.

The Senilis group of Ceraclea (Table 1) appears to have both facultative and obligate species of sponge feeding caddisflies. Larvae and adults of C. senilis and C. spongillovorax possess some morphological characteristics typical of the Fulva Group, but have not diverged markedly from other Ceraclea caddisflies. Ecologically, they appear as intermediate between the sponge feeders and non-sponge feeding species, since they can utilize additional food sources besides freshwater sponge.

The significance of immature stages and the importance of elucidating ecological relationships in developing systematic concepts can be seen from two species of Ceraclea in the Senilis Group, C. spongillovorax, and C. maculata. By examining only the male genitalia, it would appear that the male of C. spongillovorax (Fig. 1a) could be considered as being well within the morphological variation encountered in C. maculata (Fig. 2a). However, by studying the immature stages, the morphological differences in the larvae (Fig. 3 cf. Fig. 4; Fig. 6a cf. 6b) and pupae (Fig. 5a cf. Fig. 5b) indicate the possibility of two distinct species. Ecologically, the species can be separated since the larva of C. spongillovorax is a sponge feeder, while the larva of C. maculata is a detritus feeder. By rearing individual larvae to the adult stage, taxonomic characters that could be used to separate the adults were determined (Fig. 1b and 1c cf.

302

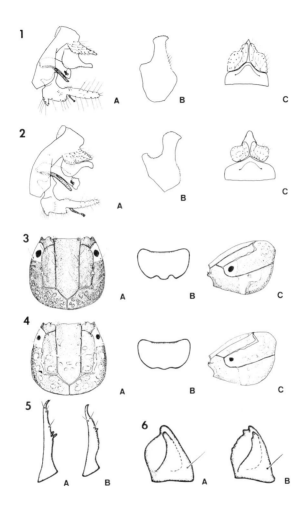

Fig. 1. Ceraclea spongillovorax (Resh) and Fig. 2. Ceraclea maculata (Banks): A. lateral view; B. tenth tergite lateral view; C. ninth and tenth tergites dorsal view. Fig. 3. C. spongillovorax and Fig. 4. C. maculata: A. larval head, dorsal view; B. gular plate; C. larval head, lateral view. Fig. 5. anal rods of pupae: A. C. spongillovorax; B. C. maculata. Fig. 6. Larval mandibles: A. C. spongillovorax; B. C. maculata.

2b and 2c). If the ecology and morphology of the immature stages of these two species had not been known, the cryptic adult stages of C. spongillovorax, would probably have remained undescribed.

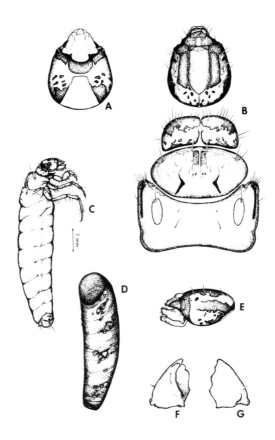

Fig. 7. Ceraclea nigronervosa (Retzuis). A. Head, ventral view; B. Head and thorax, dorsal view; C. 5th instar larva; D. Larval case; E. Head, lateral view; F. Mandible, ventral view; G. Mandible, dorsal view. Based on material collected from the Green River, near Daniel, Sublette Co., Wyoming.

The type species of Ceraclea, C. nigronervosa appears to be an obligate sponge feeder (Nielsen '48; Maitland '67; I. D. Wallace, personal communication). However, the larva of this species does have parafrontal sutures and with the exception of the extremely shortened antennae in the final larval instar and the secreted larval case, it is not markedly different from the non-sponge feeding Ceraclea. The larva of C. nigronervosa based on North American material is illustrated in Fig. 7.

Table 2. Bionomics of the freshwater sponge, Spongilla lacustris (L.) and the two cohort leptocerid caddisfly, Ceraclea transversa (Hagen) in the Salt River, Spencer Co., Kentucky. Roman numerals refer to cohort. See text for explanation.

Month	Sponge growth cm³/hr collection	Larval instar 1	2	3	4	5	Prepupa	Pupa	Adult
April	.1			II	II			I	I
May	.5		II	II	II			I	I
June	1.8	I	I	I	II	II			
July	5.0		I	I	I	II	II	II	II
August	66.3	II	II	I	I	I			
September	28.8	II	II	II	II	I	I		II
October	27.3		II	II	II	I	I		
November	6.5			II	II		I		
December	4.0			II	II		I		
January	Gemmules			II	II		I		
February	Gemmules			II	II		I		
March	Gemmules			II	II		I		

There are two basic life history patterns exhibited by sponge feeding caddisflies. In both, the life cycles are intimately linked to the bionomics of the freshwater sponge. The proliferation and growth of many species of freshwater sponge follows the pattern exhibited by Spongilla lacustris (L.) throughout most of North America. In this species, the change from the overwintering gemmule stage to the definite colonial form of the sponge begins in the spring, with peak abundance of the sponge colonies occurring in late summer and early autumn (Table 2). Lower water temperatures in autumn cause deterioration of the colonial sponge, and overwintering is again spent in the gemmule stage. The next generation of the colonial sponge will result from these gemmules.

The first type of predator life cycle exhibited by sponge feeding caddisflies is a basic univolvine pattern where only a single generation is produced in an annual cycle. A species that exhibits this pattern is Ceraclea resurgens (Walker). The adults of C. resurgens have a brief period of emergence,

flight activity, and oviposition in the spring. The eggs
are released on the water surface of streams, sink, and ad-
here to submerged stones. After hatching, the individual
larva begin case construction and search for a sponge host.
The larva of C. resurgens passes through five instars, overwinters as a prepupa, (a non-feeding fifth instar larva inside a sealed pupal case), and emerges the following spring.
Stomach analysis indicates the presence of sponge spicules
in larval instars two through five. While searching for a
sponge host, the first instar larva is probably a detritus
feeder.

In contrast, the second type of life history pattern is
represented by a population in which there are two temporally
separated periods of emergence and flight activity. A
species that exhibits this pattern is Ceraclea transversa
(Hagen). Adults of C. transversa begin emergence in spring
but flight activity and oviposition last only a few weeks.
A second period of emergence then begins in mid-summer and
extends into autumn. None of the individuals that emerge
during the first flight period participate in the second
period. Eggs, however, are released during both flight periods.

The process of hatching and the behavior of first instar
larvae of C. resurgens and C. transversa are similar. However, the population structure of the two species differs
greatly. For example, while the overwintering larval population of C. resurgens is composed of fifth instar prepupae,
individuals of the third, fourth, and fifth instar larval
stage can be found in the C. transversa population. The
difference in the range of instars for the two species represents differences in population dynamics that are due to
the presence of a two cohort cycle in C. transversa as compared to the single cohort life cycle of C. resurgens.

The first cohort of C. transversa originates from eggs
oviposited by adults during the first flight period that
occurs in early spring (Table 2). The bionomics of this cohort is essentially identical to the population dynamics of
the single cohort of C. resurgens. The larvae grow through
the summer feeding entirely on sponge, overwinter as prepupae, and then emerge the following spring.

The second cohort of C. transversa begins with eggs oviposited during a second adult flight period occurring later
in the summer (Table 2). Early instar larvae feed on sponge
but when gemmulation occurs in autumn, they must choose an
alternative food source, since the sponge gemmules are in-
edible. These third or fourth instar larvae will overwinter
as active detritus feeders, and then return to sponge feed-

ing the following spring. The adults of this second cohort will then emerge during the second adult flight period.

The two cohorts in the C. transversa population also differ in growth rates. Head widths of the fifth instar larvae of the sponge feeding first cohort measure .93 ± .03 mm. with a range from .87 - 1.01 mm. The head width of the second cohort larvae are significantly smaller, measuring .84 ± .02 mm, with a range from .73 - .91 mm. In the univoltine life cycle of C. resurgens and the first cohort of C. transversa, the sole food source is freshwater sponge. Since overwintering is spent in the non-feeding prepupal stage, the inedibility of the gemmules does not necessitate a feeding change, and in fact may serve as a trigger for the prepupal preparations. The small size of the second cohort larvae of C. transversa may be due to the high energy cost of maintaining an active overwintering larval stage, or the inefficiency of changing feeding habits and food sources.

ORDER NEUROPTERA
FAMILY SISYRIDAE
GENUS SISYRA
GENUS CLIMACIA

Spongilla flies of the family Sisyridae are commonly found in association with freshwater sponge. The life history studies and observations of spongilla flies by Needham and Betten ('01), Lampe ('11), Withycombe ('22), Old ('33), and Wesenburg-Lund ('43) are summarized in the detailed life history of Climacia areolaris (Hagen) by Brown ('52). The biology and morphology of both immature and adult stages of sisyrids have been described by Parfin and Gurney ('56) who list 17 species of spongilla flies that occur in the Western Hemisphere. Since that work originally appeared, additional distribution records and taxonomic descriptions of North American species have been reported by Acker ('61), Isom ('68), Roback ('68), Alayo ('68), Poirrier ('69), Throne ('71), Poirrier and Arceneaux ('72), and Brown ('74).

The sisyrids are the only true aquatic insects in the order Neuroptera. Although there are five described genera: Sisyra, Climacia, Sisyrina, Sisyrella, and Neurorthus, only the first is cosmopolitan in distribution. Climacia, however, is found throughout North and South America, while the latter three genera are found in the Eastern Hemisphere. The fossil record contains a sixth sisyrid genus, Rophalis, known from Baltic Amber.

Unlike the sponge feeding caddisflies, sisyrid larvae do not ingest whole particles of sponge. Their feeding behav-

ior is more typical of the Neuropterans in that spongilla
fly larvae suck fluids from the freshwater sponge. This is
possible because the mouth parts are modified into two tubes
which are formed by closely appressed maxillae and mandibles
(Parfin and Gurney, '56). Besides feeding on sponge,
spongilla flies have also been reported to suck fluids from
bryozoans and algae (Steffan, '67).

The biology of spongilla flies has been excellently described by both Brown ('52) and Parfin and Gurney ('56).
There appears to be only three larval instars, although multiple cohorts are probably present in sisyrid populations.
In Kentucky streams, all three instars of S. vicaria have
been collected in sponges from May through October, while
adult flight activity ranged from June through September.
Although second instar larvae have been collected as late as
November, spongilla flies probably overwinter as terrestrial
prepupae. However, the possibility that spongilla flies may
overwinter in other stages has not been excluded.

ORDER DIPTERA
FAMILY CHIRONOMIDAE
GENUS XENOCHIRONOMUS
GENUS DEMEIJEREA

Larval stages of chironomid species belonging to at least
a dozen different genera have been reported in association
with freshwater sponge (Weltener 1894; Lipina '27; Pagast
'34; Wesenburg-Lund '43; Wundsch '43a, '43b, '43c, '52;
Kaiser '47; Thienemann '54; Roback '68). However, only two
species, Demeijerea rufipes (L.) and Xenochironomus xenolabris (Keiffer) actually feed on the sponge (Roback '63,
'68; Maitland '66, '67).

The genus Xenochironomus contains two subgenera, Xenochironomus and Anceus (Roback '63). The former subgenus
includes X. xenolabris, the larva of which is a sponge feeder, and two species with unassociated larva , X. ugandae and
X. trisetosus. The larvae of X. xenolabris is distinct from
the other known species in that genus in that it has both a
shortened paralabial plate and a strongly arched labial
plate (Roback '63).

The taxonomic status of the genus Demeijerea has been
periodically questioned over the past half century
(Goetgheburen '28; Edwards '29; Wundsch '43a; Townes '45;
Maitland '67; Neff and Benfield '70). Morphologically, the
sponge feeding D. rufipes is distinct from other Demeijerea
species in that the eleventh post cephalic segment has only
a single pair of tubular gills, a structure that occurs in

308

only one or two other species of chironomids (Maitland '67). Roback ('68) described comparative measurements for X. xenolabris larvae and concluded that at least three larval instars were present, but that an early instar may have been missed due to the time of the year in which his collections were made. The sponge collections used in elucidating the life cycle of C. transversa in Table 2 were also used to examine the bionomics of X. xenolabris. In that population, there were four discrete larval instars, but similar to the sponge feeding caddisflies, it appears that the first instar larvae of X. xenolabris may molt soon after entering the sponge. All instars were found in freshwater sponge from June through November, with pupae and adults being collected from June through September.

Similar to X. xenolabris, the adults of D. rufipes have a long period of adult emergence (Maitland '67). It is probable that both species have multiple cohorts in an annual cycle. The stage or stages in which these sponge feeding chironomid species overwinter has not been determined.

DISCUSSION

Sponge feeding organisms are bound to a unique combination of water quality tolerances, their own and that of the host sponge. A species of sponge that can survive under a wide range of environmental conditions may support sponge feeding species only in conditions in which the predators can survive. Conversely, if the water quality tolerances of the sponge are narrower than those of the predator, neither can survive unless the predator can choose an alternative food source.

Poirrier ('69) and Poirrier and Arceneaux ('72) listed 11 species of freshwater sponge that serve as host species for spongilla flies, with seven different species that are preyed upon by Climacia areolaris. Host records for X. xenolabris include Euspongilla lacustris (Lipina '27; Pagast '34), Spongilla lacustris (Roback '63), Trochospongilla leidii (Roback '63), Spongilla fragilis (Roback '68), and Anheteromeyenia ryderi (Rio Frio, Belize, Central America, June 17, 1974. V. Resh). This last record represents the southernmost distribution reported for X. xenolabris. Ceraclea caddisflies use at least six different species of freshwater sponge as food sources.

From this information, it would appear that there is little host specificity among sponge feeders. However, the means by which a newly hatched sponge feeding larva locates a sponge host is not known. I have noticed that students in

309

ecology classes are fascinated by the distinct smells of
various species of freshwater sponge. The possibility that
sponge feeding aquatic insects may also use similar olfactory
mechanisms for locating a host sponge should be considered.
If an olfactory process is involved, is host specificity
merely a reflection of scent intensity that emanates from
the different species of freshwater sponge?

Since the life cycle of the sponge feeding species are so
closely correlated with the proliferation-gemmulation pattern in the sponge, a change in this pattern will probably
cause a modification in the predator life cycle. The pattern of proliferation and gemmulation described in Table 2
for S. lacustris is typical for most of North America. However, in some southern localities, gemmulation and sponge
proliferation occur year round. Since this modified pattern
would provide a constant food source, one would expect multiple generations of predators to be found, and in a species
with a two cohort life cycle such as C. transversa, the need
for a detritus feeding cohort would be eliminated.

The question has often been raised as to whether the sponge
feeding species are actually using the sponge as a sole food
source. Wundsch ('43c) speculated that the symbiotic green
algae found in the sponge, Chlorella, was actually the food
source of the chironomid D. rufipes. Closely related species
of the plant feeding chironomid genus Glyptotendipes behave
similarly (Gripekoven '14) and Glyptotendipes polytomus
Keiffer has been found feeding on algae while within freshwater sponge (Berg '48).

Eliminating the spongilla flies, which are more like sponge
feeding aphids than actual ingesters of sponge, this question
of algal feeding may be related to the evolution of the
sponge feeding habit. All genera that have sponge feeding
species also have systematically related species that are
algal feeders. The attraction to the sponge may have origi-
nated with an algal feeding ancestor that specialized in
feeding on the symbiotic algae found in the sponge.

Sponge obviously is not free from predators, but the total
number of species that have evolved the ability to use this
food source is still unknown. Future research on the sponge
feeding habit will probably reveal other direct associations
with freshwater sponge, such as certain species of water
mites (Pennak '53).

Studies should now begin to concentrate on the morphological and physiological basis that allows organisms to ingest
and metabolize the sponge. Jefferts et al. ('74) reported
the occurrence of two unsaturated C_{26} acids in the marine
sponge, Microciona prolifera. The presence of these long

chain fatty acids in marine and freshwater sponges could account for the few species that prey on this unusual food source. Like many other problems that need to be considered in developing a better understanding of the biology of the sponge feeding species, a search for the presence of these long chain fatty acids in either freshwater sponge or the sponge predators presents a fascinating subject for researchers to investigate.

ACKNOWLEDGMENTS

I thank Drs. Glenn B. Wiggins, Royal Ontario Museum, John C. Morse, Clemson University, Ian D. Wallace, Merseyside County Museum (Liverpool, Great Britain), Michael A. Poirrier, University of New Orleans, Selwyn S. Roback, Philadelphia Academy of Sciences, Harley P. Brown, University of Oklahoma, and Stuart E. Neff, University of Louisville, for their thoughtful comments and assistance in the development of an ecological and phylogenetic understanding of the sponge feeding habit. The research on which this report is based was supported in part by the U.S. Department of the Interior, Office of Water Resources Research, as authorized under the Water Resources Act of 1964, Project No. B-022-KY.

LITERATURE CITED

Acker, S. 1961 Report on Spongilla-flies at Clear Lake, California (Neuroptera:Sisyridae). Wasmann Jour. of Biol., 19: 283-286.
Alayo, D. 1968 Los Neuropteros de Cuba. Poeyana (Ser. B.), 2: 5-127.
Berg, K. 1948 Biological studies on the River Susaa. Folia Limnol. Scand., 4: 1-318.
Brown, H.P. 1952 The life history of Climacia areolaris (Hagen), a neuropterous 'parasite' of fresh-water sponges. Amer. Midl. Natur., 47: 130-160.
_____1974 Distribution records of Spongilla flies (Neuroptera:Sisyridae). Ent. News, 85: 31-33.
Clady, M.D. 1969 Use of freshwater sponge in case construction of Limnephilus species. (Trichoptera:Limnephilidae). Proc. ent. Soc. Wash., 71: 98.
Edwards, F.W. 1929 British non-biting Midges (Diptera, Chironomidae). Trans. Ent. Soc. London, 77: 279-430.
Goetgheburen, M. 1928 Dipteres (Nematoceres). Chironomidae III. Chironomariae. Fauna de France., 18: 1-175.
Gripekoven, H. 1914 Minierende Tendipediden. Arch. Hydrobiol. Suppl-Bd., 2: 129-130.

Hickin, N.E. 1967 Caddis larva. Larva of the British Trichoptera. Hutchinson and Co. Ltd. London. 476 p.

Isom, B.G. 1968 New distribution records for aquatic neuropterans, Sisyridae (Spongilla-flies) in the Tennessee River drainage. J. Tenn. Acad. Sci., 43: 109-110.

Jefferts E., R.W. Morales, and C. Litchfield 1974 Occurrence of cis-5, cis-9-Hexacosadienoic and cis-5, cis-9, cis-19-Hexacosatrienoic acids in the marine sponge Microciona prolifera. Lipids, 9: 244-247.

Krecker, F.H. 1920 Caddis-worms as agents in distribution of freshwater sponges. Ohio J. Sci., 20: 355.

Kaisir, E.W. 1947 Commensale og parasitiske chironomidelarver. Flora og Fauna, 53: 54-56.

Lampe, M. 1911 Beitrage zur Anatomie und Histologie der Larve von Sisyra fuscata (Diuck v. W. Rower) Fabr. Dissertation, Friedrich-Wilhelms-Universitat, Berlin. 55 p.

Lehmkuhl, D.M. 1970 A North American Trichoptera larva which feeds on freshwater sponges. (Trichoptera: Leptoceridae; Porifera: Spongillidae). Am. Midl. Nat., 84: 278-280.

Lepneva, S.G. 1964 Fauna of the U.S.S.R. Trichoptera. Larvae and pupae of the suborder Annulipalpia. (Translated from Russian, 1970.) Smithsonian Inst. and National Science Foundation, Washington, D.C. 683 pp.

Lipina, N. 1927 Die Chironomidenlarven des Oka-Bassins. Arb. Biol. Oka-Sta., 5: 37-48.

Maitland, P.S. 1966 The fauna of the River Endrick. Glasg. Univ. Publ. Stud., Loch Lomond, 2: 1-193.

_____ 1967 The larva and pupa of Demeijerea rufipes (L.) (Dipt., Chironomidae). Ent. mon. Mag., 103: 53-57.

Morse, J.C. 1975 A revision of the caddisfly genus Ceraclea (Trichoptera, Leptoceridae). Contr. Amer. Entomol. Inst., 11: 1-97.

Needham, J. and C. Betten 1901 Aquatic insects in the Adirondacks. Bull. N. York Mus., 47: 383-612.

Neff, S.E. and E.F. Benfield 1970 Notes on the status of the genus Demeijerea Krusemann. Proc. ent. Soc. Wash., 72: 126-132.

Nielsen, A. 1948 5. Trichoptera, caddisflies, with description of a new species of Hydroptila. In: Biological Studies on the River Susaa. (K. Berg, ed.) Folia Limn, Scand., 4: 123-144.

Old, M.C. 1933 Observations on the Sisyridae (Neuroptera). Papers Mich. Acad. Sci. Arts Lett., 17: 681-684.

Pagast, F. 1934 Uber die Metamorphose von Chironomus xenolabis Kieff, eines Schwammparasiten (Dipt.). Zool. Anz., 105: 155-158.

Parfin, S. and A.B. Gurney 1956 The spongilla-flies, with special reference to those of the Western Hemisphere (Sisyridae, Neuroptera). Proc. U.S.N.M., 105: 421-529.

Pennak, R.W. 1953 Freshwater invertebrates of the United States. Ronald Press Co., New York. 769 pp.

Poirrier, M.A. 1969 Some fresh-water sponge hosts of Louisiana and Texas Spongilla-flies, with new locality records. Am. Midl. Nat., 81: 573-575.

Poirrier, M.A. and Y.M. Arceneaux 1972 Studies on a southern Sisyridae (Spongilla-flies) with a key to the third-instar larvae and additional sponge-host records. Am. Midl. Nat., 88: 455-458.

Resh, V.H. and J.D. Unzicker 1975 Water quality monitoring and aquatic insects, the importance of species identifications. J. Water Poll. Contr. Fed., 47: 9-19.

Roback, S.S. 1963 The genus Xenochironomus Keiffer, taxonomy and immature stages. Trans. Amer. Ent. Soc., 88: 235-245.

_____ 1968 Insects associated with the sponge Spongilla fragilis in the Savannah River. Notulae Naturae, 412: 1-10.

Ross, H.H. 1944 The caddis flies, or Trichoptera, of Illinois. Bull. Ill. Nat. Hist. Surv., 23: 1-326.

Satija, G.R. 1964 The structure of the alimentary canal and mouth-parts of Trichoptera larvae with special reference to food and feeding habits. V. Food and feeding habits of Tinodes waeneri, Athripsodes alboguttatus, Rhyacophila dorsalis, Agapetus fuscipes, Phryganea striata, Plectrocnemia conspersa. Res. Bull. Punjab Univ., 15: 221-224.

Siltala, A.J. 1907 Zur trichopteren fauna von Savolax. Acta Soc. Fauna Flora Fenn., 29: 1-14.

Steffan, A.W. 1967 Ectosymbiosis in aquatic insects, p.207-289. In: Symbiosis. Vol. II. (S. Mark Henry, ed.). Academic Press, New York. 443 p.

Thienemann, A. 1954 Chironomus, Leben, Verbreitung und wirtschaftliche Bedleutung der Chironomiden. Binnengewaswer, 20: 1-834.

Townes, H.D., Jr. 1945 The Nearctic species of Tendipedini. (Diptera, Tendipedidae (=Chironomidae) Amer. Midl. Nat., 34: 1-206.

Throne, A.L. 1971 The Neuroptera suborder Planipennia of Wisconsin. Part II. Hemeribiidae, Polystoechotidae and Sisyridae. Mich. Ent., 4: 79-86.

Tsuda, M. and S. Kuwayama 1954 The larva of Leptocerus biwaensis which lives in freshwater sponge. Shin-konchu, 7: 12.

Withycombe, C.L.   1922   Notes on the Biology of some British
Neuroptera (Planipennia).   Trans. Ent. Soc. London, 501-
594.

Weltner , W.   1894   Anleitung zum sammeln von Susswasserschwammen nebst Bemerkungen uber die in ihnen wohnenden
Insektenlarven.   Dissertation, Berlin.

Wesenberg-Lund, C.   1943   Biologie der Susswasserinsekten.
Springer, Berlin. 682 p.

Wundsch, H.   1943a   Die in Susswasserschwammen lebenden
Dipteren larven, in besondere die Larven der Tendipediden.
Sitzber. Ges. Naturforsch. Freund Berlin, 1: 33-58.

_____1943b   Die Seen der mittleren Havel als Glyptotendi-
pesgewasser und die Metamorphose von Glyptotendipes
paripes Edwards.   Arch. Hydrobiol. Festschr. Thienemann.,
39: 362-380.

_____1943c   Die Metamorphose von Demeijerea rufipes L.
(Dipt., Tendip).   Zool. Anz., 141: 27-32.

_____1952   Tendipendidenlarven aus Susswasserschwammen in
der sammlung des Berliner zool ogischen museums.   Mitt.
Zool. Mus. Berlin., 28: 39-52.

BETTER LIVING THROUGH CHEMISTRY: THE RELATIONSHIP BETWEEN ALLELOCHEMICAL INTERACTIONS AND COMPETITIVE NETWORKS

Leo W. Buss

Department of Earth & Planetary Sciences
The Johns Hopkins University
Baltimore, Maryland   21218

and

Discovery Bay Marine Laboratory
Box 35, Discovery Bay
St. Anns, Jamaica, W.I.

ABSTRACT: The creation of high within-habitat diversity in cryptic coral reefs is found to be dependent upon the existence of specialized competitive mechanisms. Competitive networks act as diversifying processes and allelochemics as specialized competitive mechanisms. Experiments are described which support, but do not confirm, the notion that more specialized competitors can serve to increase diversity. A model is presented which defines the selective advantage to an organism possessing allelochemical interactions which allow formation of competitive networks.

An accepted generalization of modern ecology is that in competition systems, as competitors become more specialized, within-habitat diversity of the system increases. Although this notion is well established, no field demonstration of the manner in which specialized competitors can serve to increase diversity exists. This communication will provide such a demonstration for cryptic coral reef systems.

Jackson and Buss ('75) have demonstrated that allelochemical interactions (the use of toxins) probably serve as a widespread, complex and specific competitive mechanism among cryptic coral reef invertebrates. In addition, competitive networks have been identified in this system and their relation to maintenance of high diversity described (Buss and Jackson, '76). The existence of a known specialized competitive mechanism and a knowledge of the manner in which diversity is increased in the same system presents a unique opportunity to provide a case study in the topic of diversity and specialization.

I introduce briefly below competitive networks as diversifying processes and allelochemics as specialized competitive

315

mechanisms. A prediction is made as to their relationship, experiments described to test the prediction and a discussion of the coevolution of competitive networks and allelochemical interactions profered.

Competitive Networks, Allelochemics, and a Prediction

Competitive Networks as a Diversifying Process:
   Maintenance of a given level of diversity in marine, hard substrate, space-limited systems has been attributed to the interaction roles of competition, predation and disturbance. In the absence of some form of disturbance, i.e., in competition systems, both rocky intertidal and open coral reef substrates tend to become dominated by one or a few superior competitors (Dayton, '71; Paine, '66, '74; Porter, '72, '74). Disturbance, in the form of either predation or physical processes, makes space available for inferior competitors, thereby increasing diversity. This model of diversity maintenance is dependent upon a ranked hierarchy of competitive abilities (Species A greater than Species B greater than Species C), coupled with the maximum effect of disturbance operating upon the competitive dominant (Species A). According to this accepted model, predation and physical disturbance can serve to increase diversity, while competition may not.
   Jackson and Buss ('75) have suggested and demonstrated (Buss and Jackson, '76) the existence of an alternative, though complementary, model for the maintenance of diversity in space limited systems in the absence of high levels of predation and disturbance. The model requires that the overall competitive abilities of space-occupying organisms do not follow a simple linear ranked hierarchy. Such systems can be structured by competitive networks (Species A greater than Species B greater than Species C, but Species C is greater than Species A), as opposed to competitive networks (see also Gilpin, '75). The effect of competitive networks is to increase the time required for a dominant to emerge and consequently reduce the level of disturbance required to maintain a given level of diversity. Thus, where competitive networks exist, competition for space may increase diversity, and where competitive hierarchies exist, competition for space may decrease diversity.

Allelochemics as a Specialized Competitive Mechanism:
   Interspecific competition has long been recognized to be of two major types, exploitation competition and interference competition (Birch, '57; Miller, '67). The standard

dichotomy between specialized and generalized competitors is a product of the "niche theory of competition" (see Mac-Arthur, '72). Here exploitation (or utilization) of resources, e.g., prey species, is considered to primarily determine ecological segregation. Specialized competitors are those competitors that exploit a narrow range of the available resources relative to more generalized species. Most actual studies of interspecific competition, however, show that interference competition is acting and often plays the more important role in determining distribution and abundance. The dichotomy between specialized and generalized competitors, though, is easily applied to competition by interference, if one considers the actual competitive mechanisms involved (see Case and Gilpin, '75). Here specialized competitors are characterized by mechanisms which allow successful competition with a narrow range of potential competitors relative to species with more generalized mechanisms.

The following example of two competitors with different competitive mechanisms serves to illustrate the difference between specialized and generalized interference competitors as well as demonstrate the specialized nature of allelochemical interactions. Consider a system where space is limiting, predation and physical disturbance events are rare, and interference competition for space is always by physical overgrowth of a competitor. A generalized competitor would be a species characterized by a competitive mechanism designed to deal with competition with a broad range of potential competitors. For example, the ectoproct, Alcyonidum hirsutum, possesses large spines bordering the colony margin (see Hayward, '73). Such spines serve as an effective interference competitive mechanism against all species which cannot overgrow this ectoproct without avoiding the spines. In contrast, a specialized competitor would be a species characterized by a competitive mechanism designed to deal with a narrow range of potential competitors. For example, the encrusting sponge, Toxemna sp., produces a species-specific toxin which allows overgrowth of the ectoproct Steganoporella magnilibris which it could not, presumably, overgrow without killing (see Jackson and Buss, '75). Thus the ectoproct possessing the spines can be considered a generalized competitor relative to the sponge because the spines represent a mechanism designed for competition with a broad range of potential competitors, whereas allelochemicals represent a mechanism designed for competition with only one or a few potential competitors.

A Prediction

Given the above introduction to competitive networks as
diversifying processes and allelochemical interactions as
specialized competitive mechanisms, I can now profer a prediction as to the manner in which they are related in cryptic coral reef systems.
Cryptic coral reef environments of caves, crevices, and
the under-surfaces of foliaceous corals comprise a major
habitat type of coral reefs (Hartman and Goreau, '70; Jackson, et al., '71; Lang, '74; Vacelet and Vasseur, '66; Vasseur, '64). Cryptic environmental surfaces are almost entirely covered with sessile animals, free space is almost
entirely lacking, and competition for space is intense.
Predation or signs of predation (tooth marks, drill holes,
bare zooids, etc.) are extremely rare and physical disturbance is unlikely in cryptic environments. In addition,
available substratum is found only in patches of varying
size, such that only a small fraction of the entire community
will occupy any given patch at a particular time. Encrusting sponges, sclerosponges (Hartman and Goreau, '70) and
bryozoans dominate the fauna along with numerous ahermatypic
corals, brachiopods, bivalve molluscs, ascideans, colonial
foraminifera, coralline algae, and serpulids.
Diversity of the entire system is dependent upon and
directly related to diversity on the various patches (Jackson and Buss, ms.). I will, therefore, limit discussion of
cryptic coral reef diversity to within-patch diversity. Consider any one patch occupied by, say, four species, such that
the colony margin of any one species will contact the margins
of all other species. The interference competitive abilities
of the four species may form either a competitive hierarchy
or a competitive network (Fig. 1) As demonstrated above, if
a competitive network exists the diversity on this patch will
always be greater than or equal to the diversity if a competitive hierarchy exists (Buss and Jackson, '76). Furthermore, the existence of a competitive network is clearly dependent on the outcome of the Species A versus Species D
interaction (Fig. 1). Within-patch diversity, then, is
determined by whether or not a competitive network exists
on the patch, which is, in turn, determined by the outcome
of one competitive interaction.
Having reduced for cryptic coral reefs the question of
high within-patch diversity to the result of a particular
competitive interaction, a prediction can be made. Conven-
tional wisdom on the topic of diversity and specialization
(e.g., MacArthur and Wilson, '67; MacArthur, '72) dictates

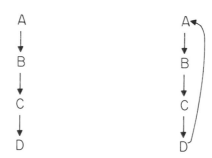

COMPETITIVE HIERARCHY   COMPETITIVE NETWORK

Fig. I A comparison of competitive networks and competitive hierarchies. Unless reverse arrows occur, any species presented below another is the loser in competition with the species above. Arrows point from the dominant of a two species interaction to the inferior competitor.

that a specialized competitive mechanism should serve to increase diversity in a manner in which a generalized mechanism could not. I predict, on the basis of this generalization, that upon investigation of competitive interactions which allow formation of competitive networks (and, therefore, allow maintenance of high within-patch diversity), that specialized competitive mechanisms will be found. Furthermore, on the basis of previous investigations into the interference competitive mechanisms of cryptic coral reef invertebrates (Jackson and Buss, '75), I predict that the mechanism of this interaction will be found to be allelochemical in nature.

Presented below are experiments designed to test these predictions.

METHODS

Coral plates were collected for study of cryptic organism interactions (Jackson et al., '71; Hartman and Goreau, '70; Jackson and Buss, '75). Nineteen cryptofaunal species were chosen for examination of evidence for competitive networks on the basis of their availability. These included eight ectoproct species, seven sponge species, one colonial a-hermatypic coral, one coralline alga, one colonial foraminiferan, and one ascidian species. Colonies of foliaceous

corals (<u>Agaracia</u> and <u>Monastrea</u> sp.) were collected using
SCUBA during July and August, 1975 from depths of -15 to -30
meters at Discovery Bay and Rio Bueno along the north coast
of Jamaica, W.I. The undersurface of each coral was examined microscopically for contacts between colony margins
of those species included in the study. Since competitive
superiority is measured by the ability to physically overgrow
a potential competitor, the dominant of each two-species
contact is readily discernible. Collection and examination
of coral plates was continued until a complete matrix of all
possible two-species contacts was obtained for several
species. Tabulation of any subset of this matrix allows
determination of whether a competitive network or a competitive hierarchy exists among the species of that subset.

The potential of allelochemical interactions that form competitive networks was evaluated by subjecting ectoprocts to
whole organism homogenates of various sponges and comparing
survival and feeding activities relative to controls after
the method of Jackson and Buss ('75).

RESULTS

A. Competitive Networks

Results are summarized in Table 1. Identification of many
organisms is lacking, due to the extraordinary taxonomic
difficulties presented by the reef cryptofauna. Complete
identifications will be available from the author upon re-
quest. Several two-species interactions tend to produce
either of the two possible results, e.g., <u>Stylopoma</u> <u>spongi-</u>
<u>ties</u> may overgrow <u>Reptadeonella</u> <u>hastingsae</u>, but the reverse
is often the case. This situation seems to be largely limit-
ed to ectoproct interactions and appears to result from complications caused by the spatial heterogenity of the environment (Buss, unpublished data). Some two-species contacts
were not obtained and this information is presently unavailable. Nevertheless, competitive networks clearly exist, and
are both numerous and complex. A subset of the entire network is illustrated in fig. 2.

B. Allelochemical Experiments

Results are summarized in Table 2. All of the sponge
species tested displayed an allelochemical effect. All
ectoproct species suffered some mortality. In cases of
adverse effects, mortality or cessation of movement and
feeding occurred in all replicates. Controls suffered no
mortality or cessation of feeding in all replicates. Allelochemical effects seem quite specific among ectoprocts

CONTACT DATA MATRIX

	B1	B2	B3	B4	B5	B6	B7	B8	S1	S2	S3	S4	S5	S6	S7	Cl	F1	C1	A1
B-1		←	←	←	←	←	⇕	←	⇕	⇕	⇕	⇕	↑	←	←	⇕	⇕	⇕	↑
B-2			⇕		⇕	↑	↑	⇕	↑	↑	↑	↑	↑	↑	⇕	↑	⇕	↑	↑
B-3				⇕	⇕	↑	←	↑	↑	↑	↑	↑	↑	⇕	↑	↑	⇕	⇕	↑
B-4										⇕	↑			←		←	↑		
B-5									⇕	←	↑	↑	↑	↑	⇕		⇕	↑	↑
B-6							↑			↑	↑	↑	↑			↑	↑		↑
B-7								←			↑	←	←	←	←	↑			↑
B-8										↑	↑	↑	←	↑			⇕	↑	↑
S-1										↑	↑	↑	↑	↑			⇕	⇕	
S-2											←	←	↑	←	↑	←	←	↑	↑
S-3												↑	←	←	↑	←	↑	←	↑
S-4													←	←	←	←	⇕	←	←
S-5														←	↑	←	↑	↑	↑
S-6															↑	↑	↑	←	↑
S-7																←	↑	←	↑
F-1																		↑	↑
C-1																			↑
A-1																			

Table I. Data from microscopic examination of competitive interactions. Arrow points toward the dominant of a two species interaction. Double arrows indicate that either result is possible. Species with the prefix B are bryozoans, S, sponges, Cl, coral, F, foraminifera, C, coralline algae, A, ascidean.

as no sponge species caused mortality in all ectoprocts tested.

DISCUSSION

The experiments provide strong circumstantial evidence for specific allelochemical interactions between three sponge and ectoproct species. Data from the contact matrix (Table 1) shows that many competitive networks are formed from these interactions (see Table 3). Thus, the results confirm the prediction that allelochemical interactions should be found in situations where competitive networks are formed and

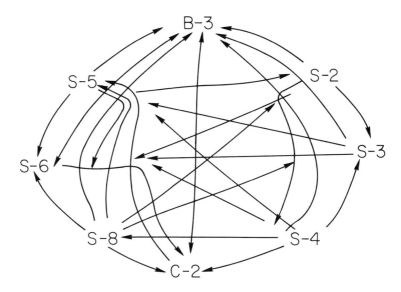

Fig. 2   A competitive network formed by compilation of
a subset of the data presented in Table I.

ALLELOCHEMICAL INTERACTIONS AND
NUMBER OF NETWORKS FORMED

Allelochemical Interaction	# Species in Network	#3-Species Network	#4-Species Network
S-2 vs B- I	16	2	84
S-3 vs B- 2	17	2	133
S-7 vs B- 3	13	2	121

Table 3. Number of 3 and 4 species networks
formed by interactions which were found to be
allelochemical in nature.

support the generalization that as competitors become more specialized, within-habitat diversity increases.

The significance of this result, however, to the niche theory generalization on diversity and specialization should not be overemphasized. Niche theories of competition (e.g., MacArthur, '72) have been repeatedly demonstrated by field ecologists to have the unfortunate tendency of producing the "right prediction for the wrong reason" (Dayton, '73, '74; Wilbur, et al., '74; Jackson and Buss, ms.). Such theories are based on assumptions, incorporated for mathematical simplicity, which are rarely, if ever, met in nature. As a result they tend to be inherently untestable. This work, therefore, should not be considered a test of any particular theory, but rather it is a case study of the relationship between diversity and competitive specialization which lends support to the niche theory result.

Our understanding of relationships of a general nature such as that of diversity and specialization is dependent upon appreciation of the manner in which these relationships have evolved. Basic to this understanding is a knowledge of the selective advantage to the organism in producing the result that supports the generalization. Clearly, there is no advantage to an organism in increasing within-habitat diversity. That is to say, there is no evidence to support selection of organisms on the basis of a community attribute (e.g., diversity, complexity, etc.). Although development of a specialized competitive mechanism may tend to increase diversity, the mechanism is not selected for because it can increase diversity, but because the organism's strategy for survival is dependent upon it. Thus, increased diversity must be considered an incidental result.

For the system described here, the evolutionary question becomes: How do those organisms with specific allelochemical interactions which form diversity-increasing competitive networks benefit from the development of such a mechanism. To answer this we must again consider the nature of the system in question. Space is limiting and substratum is found in patches of varying size (Buss and Jackson, '76). Survival of these organisms is dependent on their maintaining a stable (MacArthur, '72), although not necessarily large (Slobodkin, '68) population on some fraction of the available patches. If an organism can maintain single or few species dominance on a sufficient number of the available patches as to guarantee larval production sufficient to colonize newly created patches, such a stable population could be achieved (Dayton, '74; Levin and Paine, '74; MacArthur, '72). Thus if a demonstration is provided of the manner in which an allelochemical competitive mechanism can allow single or few species dominance on some fraction of the available substrata, then the selective advantage to the organism possessing this mechanism is sufficiently demonstrated.

How, then, can possession of allelochemicals by an encrusting sponge serve to guarantee dominance or near dominance on some fraction of the available patches? I will first demonstrate the manner in which single species dominance can be achieved by such organisms on a single patch and then provide evidence that the number of patches where single species dominance may be achieved is large. Consider figure 3. Assume a four species network as illustrated with the Species A versus Species D interaction being allelochemical in nature (Fig. 3a). Figure 3b is a schematic representation of the surface area covered on a single patch, by the four species at some time t_0. Assume that (1) the four species are positioned as illustrated, (2) Species A and Species D have comparable growth rates, and (3) Species B and Species C have growth rates less than those of Species A or D. Then at some time t_1 this patch will become completely dominated by Species D (Fig. 3c). Note that on this patch, two of the three potential competitor species are competitively superior to Species D, but that by possession of a specific allelochemical interaction that allows formation of a network, Species D is able to achieve dominance.

ALLELOCHEMICAL EXPERIMENTS

Homogenates	Species Exposed to Homogenates		
	Steganoporella magnilibris (B-1)	*Stylopoma spongites* (B-2)	*Reptadeonella violacea* (B-3)
S-2	NTE	NTE	D
S-3	NMF	D	NTE
S-7	D	NTE	NTE

Table 2. Results from addition of sponge homogenates to aquaria containing ectoprocts. NTE (no apparent toxic effect ; normal movement and feeding of ectoproct zooid); NMF (no movement or feeding ; zooids intact); D (ectoproct colonies dead)

The question remains as to whether dominance can be a-
chieved in this manner on a sufficient number of patches to
allow maintenance of a stable population structure. Table 3
represents a compilation of the number of three and four
species networks formed by the three sponges demonstrated
to possess allelochemical competitive mechanisms. This
table includes only networks for which contact data was
available and, therefore, certainly underestimates the
actual number of networks formed. Furthermore, as more
species are added to the network (5-species networks and
above) the number of possible combinations which may result
in dominance increases. Clearly, the possession of an allelochemical competitive mechanism by these encrusting
sponges allows formation of not one but numerous competitive networks, each of which may result in single species
dominance.

In summary, discussion of the coevolution of competitive
networks and allelochemical interactions requires identification of the selective advantage to the organism possessing
this competitive mechanism. Given the nature of the system
in question, the organisms must maintain a stable population
on some fraction of the available substrate patches to
survive. Single or few species dominance of patches can
lead to such stable population structure. It is demonstrated
that allelochemical interactions which allow formation of
competitive networks can also lead to dominance on single
patches and that the number of patches where this result is
possible is large. Consequently, there is likely to be
strong selection for development of specific allelochemical
interactions which allow formation of competitive networks.

ACKNOWLEDGMENTS

Brian D. Keller was involved in the collection of all
specimens for both the determination of competitive networks
and the allelochemical experiments. Without his participation this project could not have been attempted. Additional
diving assistance was provided by J.B.C. Jackson and D.W.
Wethey. J.B.C. Jackson, B.D. Keller, J.C. Lang, and K.
Rylaarsdam read the manuscript, all providing useful comments. Financial support was provided by the National
Science Foundation (Grant # DES72-01559-A01) and Sigma Xi.

LITERATURE CITED

Birch, L.C.   1957 The meaning of competition. Amer. Natur.,
    91: 5-18.

Buss, L.W. and J.B.C. Jackson 1976 Interference competitive networks: the maintenance of diversity in marine hard-substrate environments. Submitted to Science.

Case, T.J. and M.E. Gilpin 1974 Interference competition and niche theory. Proc. Nat. Acad. Sci., U.S.A., 71: 3073-3077.

Dayton, P.K. 1971 Competition, disturbance and community organization: the provision and subsequent utilization of space in a rocky intertidal community. Ecol. Monogr., 41: 351-389.

_____1973 Two cases of resource partitioning in an intertidal community: making the right prediction for the wrong reason. Amer. Natur., 107: 662-670.

_____1974. Dispersion,dispersal, and persistance of the intertidal alga, Postelsia palmaeformis Ruprecht. Ecology, 54: 433-438.

Gilpin, M.E. 1975 Limit cycles in competition communities. Amer. Natur., 109: 51-60.

Hartman, W.D. and T.F. Goreau 1970 Jamaican coralline sponges: their morphology, ecology and fossil relatives. Symp. Zool. Soc. Lond., 25: 205-243.

Hayward, P.J. 1973 Preliminary observations on settlement and growth of Alcyonidium hirsutum (Fleming). Pages 107-113 in G.P. Larwood, ed. Fossil and Living Bryozoa. Academic Press, London.

Jackson, J.B.C. and L.W. Buss 1975 Allelopathy and spatial competition among coral reef invertebrates. Proc. Nat. Acad. Sci., U.S.A., in press.

Jackson, J.B.C., T.F. Goreau, and W.D. Hartman 1971 Recent brachiopod-corraline sponge communities and their paleoecological significance. Science, 173: 623-625.

Lang, J.C. 1974 Biological zonation at the base of a reef. Amer. Scien., 62: 272-281.

Levin, S.A. and R.T. Paine 1974 Disturbance, patch formation and community structure. Proc. Nat. Acad. Sci., U.S.A., 71: 2744-2747.

MacArthur, R.H. 1972 Geographical Ecology. Harper and Row, N.Y.

Miller, R.S. 1967 Pattern and process in competition. Adv. Ecol. Res., 4: 1-74.

Paine, R.T. 1966 Food web complexity and species diversity. Amer. Natur., 100: 65-75.

_____1974 Intertidal community structure: experimental studies on the relationship between a dominant competitor and its principal predator. Oecologia, 15: 93-120.

Porter, J.M. 1972 Predation by Acanthaster and its effect on coral species diversity. Amer. Natur., 106: 487-492.

_____1974 Community structure of coral reefs on opposite sides of the Isthmus of Panama. Science, 186: 543-545.

Slobodkin, L.B. 1968 Toward a predictive theory of evolution; pages 317-340 in Lewontin,R., ed., Population Biology and Evolution, Syracuse Univ. Press.

Vacelet, J. and P. Vasseur 1966 Les tunnels obscurs sous-recifaux de Tuléar et leur faune de spongiatres. 2nd Intern. Conf. Oceanogr. Moscow, 443: 378.

Vasseur, P. 1964 Contribution a l'étude bionomique des peuplements sciaphiles infralittoraux de substrat dur dans les recifs de Tuléar, Madagascar. Rec. Trans. Stat. Mar. Endoume, fasc. hors. serie supple., 2: 1-72.

Wilbur, H.M., D.W. Tinkle, and J.P. Collins 1974 Environmental certainty, trophic level, and resource availability in life history evolution. Amer. Natur., 108: 805-817.

A NEW CERATOPORELLID SPONGE (PORIFERA:SCLEROSPONGIAE)
FROM THE PACIFIC

Willard D. Hartman

Department of Biology
and
Peabody Museum of Natural History
Yale University
New Haven, Connecticut 06520

and

Thomas F. Goreau[1]

Discovery Bay Marine Laboratory,
University of the West Indies
and
State University of New York at Stony Brook

ABSTRACT: A ceratoporellid sponge, Stromatospongia
micronesica sp. nov., is described from underwater caves,
tunnels and crevices in reef environments of the Marianas
and Marshall Islands. The sponge is distinguished from
its Caribbean relatives by the more variable form of the
surface processes of the basal aragonitic skeleton and by
the smaller size of the siliceous spicules.

The order Ceratoporellida was proposed (Hartman and Goreau,
'72) to include four genera of Recent sclerosponges with a
compound skeleton of aragonite, proteinaceous fibers and
spicules of silicon dioxide. Cuif ('73, '74) has described
Triassic representatives of the order, including an extinct
species of Ceratoporella as well as two new genera, from the
Dolomite Alps of northern Italy. The Recent ceratoporellids
so far described all dwell in cryptic environments in the

---

[1]Thomas F. Goreau, who was professor of marine sciences at
the University of the West Indies, Mona, Kingston, Jamaica,
professor of biology at the State University of New York
at Stony Brook and director of the Discovery Bay Marine
Laboratory, Discovery Bay, Jamaica, died unexpectedly in
New York on April 22, 1970. This paper, based in part on
specimens collected by Professor Goreau and enriched by his
observations of the organism in life, was prepared by the
first-named author.

tropical western Atlantic. It is the purpose of this paper to describe a species from similar habitats in the western Pacific region.

METHODS

Specimens, collected by SCUBA diving, were fixed in 10% neutral formalin as soon as possible after being brought to the surface. For long term storage the specimens were transferred to 75% ethanol after a few days to several months in formalin. Tissue-free calcareous skeleton preparations were obtained by treatment with 5.25% sodium hypochlorite solution (commercial bleach). Siliceous spicules were isolated from organic material and calcareous skeleton by treatment with concentrated nitric acid (see Hartman, '75, for further details). Scanning electron micrographs were prepared from platinum- or gold-coated specimens on one of the following instruments: ETEC Autoscan U-1, Cambridge Stereoscan and JEOL JSM-U3. Histological studies were carried out on formalin-fixed material that was decalcified in 2% formic acid and stained in Heidenhain's iron hematoxylin and Alcian blue.

DESCRIPTION

Stromatospongia micronesica sp. nov.

DIAGNOSIS. Basal skeleton aragonitic, spreading over the substratum, seldom exceeding 1 cm in thickness. Surface of basal skeleton produced into closely spaced multibranched or lamellate processes up to 0.9 mm high, the living tissue extending into the spaces between the processes. Siliceous spicules are acanthostyles ornamented with whorls of spines. Range of means of length and width of the spicules are: 112 to 130 µm x 5.5 to 6.0 µm (8 specimens).

THE CALCAREOUS SKELETON. The basal calcareous skeleton of S. micronesica is composed of fibrous aragonite and among populations from the Marianas Islands varies in thickness from 1.5 to 8.5 mm. The sponge overgrows calcareous reef rock (Fig. 1) or other sponges with a basal calcareous skeleton (Fig. 2) such as Astrosclera sp. or Acanthochaetetes wellsi Hartman and Goreau ('75). A fragment of a specimen from Enewetak Atoll suggests that the entire skeleton might have been 3 cm thick but it is difficult to be certain of its thickness because of the angle of fracture of the fragment from the piece left in the field. A single continuous specimen may cover rounded and irregular surfaces that spread over a plane area of at least 100 cm^2.

The surface of the calcareous skeleton is produced into multibranched to lamellate processes, the configurations varying from specimen to specimen among the Guam populations (see Figs. 3, 4, 5, 6). Specimens from Saipan have finely branched processes and show little variation within the population according to the material available for study. The irregular spaces between the processes extend to depths of 0.7 to 0.9 mm and are penetrated by the living tissue of the sponge. These spaces are gradually filled in with secondary deposits of aragonite so that the basal region of the skeleton is composed of solid aragonite.

The microstructural units of the calcareous skeleton are sclerodermites each formed of acicular crystalline units of aragonite radiating in all directions from a center of calcification (see Hartman, '69; Barnes, '70; Cuif, '74). Progressive enlargements of the acicular crystalline units of aragonite are portrayed in figures 7, 8, 9, 10. Figure 10 gives evidence of the existance of subunits making up the acicular crystalline units of the sclerodermites.

SILICEOUS AND ORGANIC SKELETAL ELEMENTS

The siliceous spicules are acanthostyles ornamented with whorls of spines (Figs. 11, 12, 13). That end of the spicule with the greater diameter may terminate in one or two rounded projections, forked or not, and up to 3 μm or more in height, or the end may be gently rounded and lacking in terminal projections. There follows a whorl of straight, blunt, rounded projections, that are sometimes forked. Then follow two or three whorls of sharply pointed spines recurved toward the pointed end of the spicule; next there occur one or two whorls of straight, pointed spines of lesser height and finally a series of whorls of spines recurved toward the end of the spicule of greater diameter. At first the spines of these whorls decrease in height to the mid-region of the spicule where they again progressively increase in height until they become prominent hooks at the pointed end of the spicule. Many variations occur. The height of spines making up the whorls of the mid-region of the spicule may differ little from that of the ends (Fig. 11). The height of the spines distal to the whorls at the rounded end of the spicule may be uniformly low (Fig. 13), not increasing at the pointed end. The distal ends of the styles may be truncate (Fig. 11).

The range of means (with standard error) and overall ranges of lengths and widths of acanthostyles of specimens from the several localities are as follows:

GUAM, 4 specimens, 30 measurements per specimen:
  Length, 112 (±3) to 130 (±4) μm; overall range, 64-197 μm.
  Width, 5.5 (±0.2) to 6.0 (±0.2) μm; overall range, 2.7-7.8
  μm.
SAIPAN, 4 specimens, 30 measurements per specimen:
  Length, 117 (±3) to 129 (±4) μm; overall range, 78-189 μm.
  Width, 4.6 (±0.1) to 5.9 (±0.1) μm; overall range, 2.3-7.2
  μm.
ENEWETAK, 1 specimen, 30 measurements per specimen:
  Length, 116 (±5) μm; overall range, 78-174 μm.
  Width, 7.2 (±0.3) μm; overall range, 3.6-10.4 μm.

As the basal calcareous skeleton grows upward the spicules
may become enclosed in aragonite (Fig. 14) and often protrude
from the surface of this skeleton (Figs. 4, 6). Those siliceous spicules that are embedded in aragonite often become
partly eroded (Fig. 12, two spicules on right). The rounded
ends of the siliceous spicules are enclosed in spongin which
forms a network in the living tissue of the sponge.

LIVING TISSUES OF THE SPONGE

The living tissue of S. micronesica varies in color from
cream to yellowish cream to yellow-tan to ochre. Specimens
from the Saipan population tend to be pinkish maroon when
growing near the entrances to caves or tunnels. Guam specimens are occasionally splotched or tinged with maroon or rose
when growing in less subdued light. Most specimens from
Guam manifest a light suffusion of olive green or gray that
may relate to the presence of boring algae in the calcareous
skeleton. The surface of the living tissue is marked by a
reticulum of exhalant channels formed by the fusion of adjacent channel systems each composed of an oscule and the
channels that radiate toward it (Fig. 15). The oscules vary
from 260 to 360 μm in diameter and lie from 5 to 10 mm apart
on the surface of the sponge. Ostia lie in the spaces between the reticulum of excurrent channels and vary from 50 to
70 μm in diameter in the expanded state. All these measurements are based on close-up photographs of living sponges in
their cave habitats. The choanocyte chambers range from 12
to 16 μm in diameter as measured in histological preparations.
  SYMBIONTS. Although Caribbean species of Stromatospongia
show an obligatory or frequent association with serpulid
polychaetes of the genus Pseudovermilia (see Hartman, '69;
ten Hove, '75), such occurrences are rare among the populations of S. micronesica. The brachiopod Thecidellina congre-

gata Cooper is a frequent associate of the Pacific sclero-
sponge in question, however (Jackson et al., '71). Fila-
mentous boring algae are of common occurrence in the surface
processes of the calcareous skeleton.

RANGE AND HABITAT

Stromatospongia micronesica is known at present from Guam
and Saipan, Marianas Islands, and from Enewetak Atoll,
Marshall Islands. In Guam it is known from two sites on the
west coast: on the walls of an underwater cave on Anae Island with the entrance at 4.5 m and the floor at 9 to 11 m
and in deep recesses in a blue hole south of Orote Point at
a depth of 40 m, the greatest depth at which it is known at
present. In Saipan it has been found encrusting the walls
of an underwater tunnel and grotto at Puntan Madog at depths
of 3 to 6 m. This species tends to extend its range into the
darkest recesses of caves and tunnels where Astrosclera and
Acanthochaetetes wellsi do not occur.

HOLOTYPE. Peabody Museum of Natural History, Yale University, YPM No. 9104 (Fig. 2). Cave on Anae Is., Guam, 6 m.
Collected by T.F. Goreau, August 9, 1969.

Repositories of paratypical material: National Museum of
Natural History, Smithsonian Institution, Washington, D.C.;
British Museum (Natural History), London; Peabody Museum of
Natural History, Yale University. About twenty lots were
studied.

COMPARISON WITH OTHER SPECIES OF STROMATOSPONGIA

S. micronesica is readily separable from the West Indian
species, S. vermicola and S. norae, on the basis of the much
smaller siliceous spicules present in the Pacific form.
Comparative measurements follow as ranges of means:

S. micronesica, 9 specimens, 30 measurements per specimen:
  Length, 112 (±3) to 130 (±4) μm.
  Width, 4.6 (±0.1) to 7.2 (±0.3) μm.
S. norae, 3 specimens, 50 measurements per specimen:
  Length, 195 (±11) to 215 (±10) μm.
  Width, 5.5 (±0.1) to 6.1 (±0.2) μm.
S. vermicola, 3 specimens, 100 measurements per specimen:
  Length, 165 (±7) to 187 (±8) μm.
  Width, 6.2 (±0.2) to 8.0 (±0.2) μm.

The surface processes of the calcareous skeleton of S.
micronesica are usually similar in general form to those of

$\underline{S}$. $\underline{vermicola}$ except that the surface ornamentation of the skeleton of the former sponge is more delicate, the processes reaching heights up to only 0.9 mm above the solid aragonitic base as compared to 2.0 mm for the West Indian species. The lamellate surface processes that occur in some specimens of $\underline{S}$. $\underline{micronesica}$ are similar in size and form to those found on the surface of the calcareous skeleton of $\underline{S}$. $\underline{norae}$, a species that is readily separable from the Pacific form on the basis of spicule size.

$\underline{S}$. $\underline{micronesica}$ spreads rapidly over the substrate and in this characteristic resembles $\underline{S}$. $\underline{vermicola}$. Since the Pacific sponge frequently encrusts the dead rounded skeletons of other sclerosponges, it assumes a lobate form and appears to be massive.

REMARKS. The specific name refers to the occurrence of the sponge in Micronesia.

ACKNOWLEDGMENTS

H.M. Reiswig worked with one of us (W.D.H.) in the Marianas Islands in 1971 and helped immeasurable in the study of this sponge. The director and staff of the Marine Laboratory, University of Guam, offered us their generous hospitality during our work there. J.P. Villagomez assisted us greatly during our visit to Saipan. R.A. Kinzie III provided a specimen of the sponge from Enewetak Atoll.

The scanning electron micrographs were prepared with the skilled collaboration of W. Brown, J.W. Rue, V. Peters and A.S. Pooley. G. Chaplin, L.S. Keller and W. Phelps helped in technical and photographic work, and C. Dean typed the manuscript. To all these persons we express our gratitude.

The first named author owes a special debt of gratitude to Mrs. Nora I. Goreau who generously made available to him for study the collections, photographs and notes assembled by her husband before his untimely death.

This work was supported in part by a grant from the National Geographic Society.

LITERATURE CITED

Barnes, D.J. 1970 Coral skeletons: an explanation of their growth and structure. Science, $\underline{170}$: 1305-1308.

Cuif, Jean-Pierre 1973 Mise en évidence des premières Sclérosponges fossiles, dans le Trias des Dolomites. C.R. Acad. Sci. Paris, $\underline{277}$, sér. D: 2333-2336.

_____1974 Rôle des Sclérosponges dans la faune récifale du Trias des Dolomites (Italie du Nord). Geobios,$\underline{7}$(2):139-153.

Hartman, W.D.   1969   New genera and species of coralline
    sponges (Porifera) from Jamaica.   Postilla, 137: 39p.
    _____1975 Phylum Porifera.   In: Light's Manual: Intertidal
    Invertebrates of the Central California Coast.   3rd ed.
    R.I. Smith and J.R. Carlton, eds.   University of California
    Press, Berkeley, p. 32-64.
Hartman, W.D. and T.F. Goreau   1972   Ceratoporella (Porifera:
    Sclerospongiae) and the chaetetid "corals".   Trans. Conn.
    Acad. Arts Sci., 44: 131-148.
    _____1975 A Pacific tabulate sponge, living representative
    of a new order of sclerosponges.   Postilla, 167: 21 p.
Jackson, J.B.C., T.F. Goreau and W.D. Hartman   1971   Recent
    brachiopod-coralline sponge communities and their paleo-
    ecological significance.   Science, 173: 623-625.
ten Hove, H.A.   1975   Serpulinae (Polychaeta) from the
    Caribbean: III - The genus Pseudovermilia.   Stud. Fauna of
    Curaçao and other Caribbean Islands, 47: 46-101.

Fig. 1.   Upper surface of flattened specimen of Stromato-
spongia micronesica sp. nov. overgrowing reef rock.   Cave,
Anae Is., Guam.   Collected by T.F. Goreau.   YPM No. 9105.
Paratype.   Scale = 1 cm.

Fig. 2.   Upper surface of pseudolobate specimen of S. micronesica sp. nov. overgrowing dead skeletons of Acanthochaetetes wellsi Hartman and Goreau.   Cave, Anae Is., Guam.   Collected by T.F. Goreau.   YPM No. 9104.   Holotype.   Scale = 1cm.

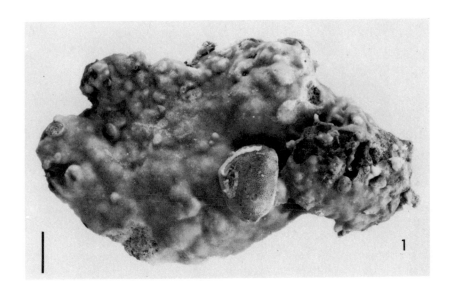

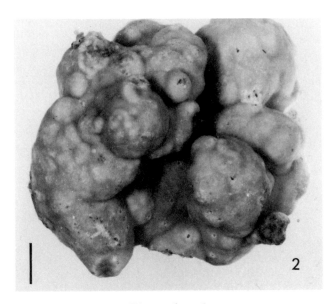

Figs. 1 & 2

Fig. 3.   Branching surface processes of basal aragonitic skeleton of S. micronesica sp. nov. from Guam.   YPM No. 9106. Scanning electron micrograph.   x60.   Scale = 0.2 mm.

Fig. 4.   Enlargement of portion of previous photograph showing aragonitic processes with protruding siliceous spicules. SEM.   x150.   Scale = 0.1 mm.

Figs. 3 & 4

Fig. 5. Lamellate surface processes of basal aragonitic skeleton of S. micronesica sp. nov. from Guam. YPM No. 9107. SEM. x65. Scale = 0.2 mm.

Fig. 6. Enlargement of portion of previous photograph showing aragonitic processes with protruding siliceous spicules. SEM. x150. Scale = 0.1 mm.

Figs. 5 & 6

Fig. 7.   Fractured surface of process of S. micronesica sp.
nov. showing crystalline units of aragonite.   YPM No. 9104.
SEM x1200.   Scale = 0.01 mm.

Fig. 8.   Fractured surface of process of S. micronesica sp.
nov. showing crystalline units of aragonite.   YPM No. 9107.
SEM x2900.   Scale = 0.005 mm.

Fig. 9.   Enlargement of crystalline units of aragonitic basal
skeleton of S. micronesica sp. nov. YPM No. 9107.   SEM.
x6000.   Scale = 0.002 mm.

Fig. 10.   Enlargement of crystalline units of aragonitic
basal skeleton of S. micronesica sp. nov. showing subunits
of crystalline structure.   YPM No. 9104.   SEM. x24,000.
Scale = 0.0005 mm.

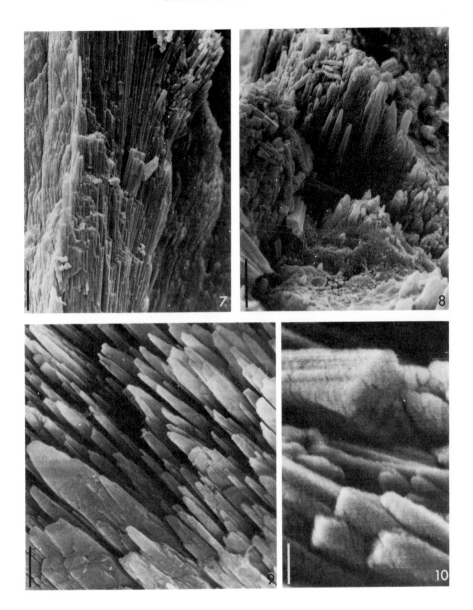

Figs. 7-10

Fig. 11. Siliceous spicules of S. micronesica sp. nov. Specimen from Guam with truncate acanthostyles. YPM No. 9108. SEM. x700. Scale = 0.02 mm.

Fig. 12. Siliceous spicules of S. micronesica sp. nov. Specimen from Guam showing two eroded spicules on right. YPM No. 9106. SEM. x700. Scale = 0.02 mm.

Fig. 13. Siliceous spicules of S. micronesica sp. nov. Specimen from Saipan. YPM no. 9109. SEM. x700. Scale = 0.02 mm.

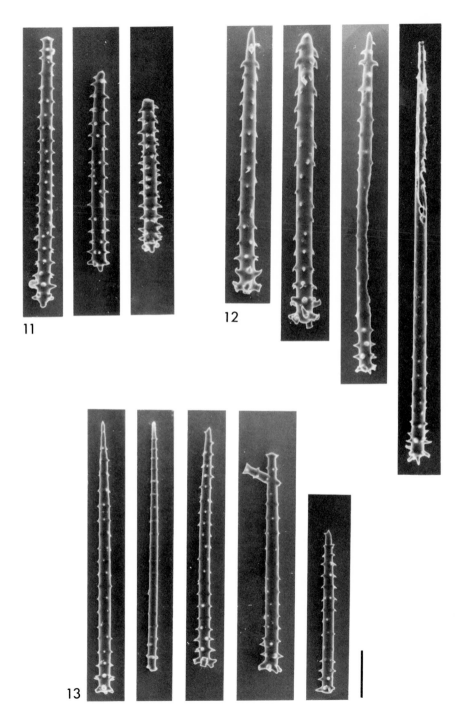

11

12

13

Fig. 14. Tips of surface processes of basal aragonitic skeleton of S. micronesica sp. nov. showing overgrowth of siliceous spicules by calcareous material. Sepcimen from Saipan. YPM No. 9110. SEM. x600. Scale = 0.02 mm.

Fig. 15. Surface features of living specimen of S. micronesica sp. nov. photographed in situ in cave at Anae Is., Guam. Note reticulum of excurrent channels, oscules and ostia. x13.25. Scale = 1 mm. Photo by H.M. Reiswig.

14

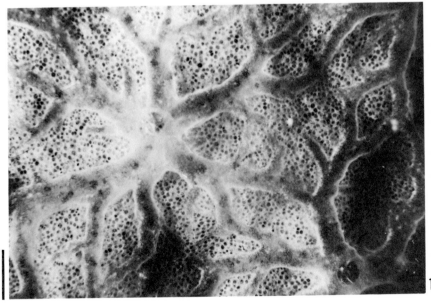

15

Subject Index

Because of the considerable number of species in the distribution lists found in Hechtel, G.J., "Zoogeography of Brazilian Marine Demospongiae", species references to listed material in that paper are not included in this index.